Gravitational Waves

Ajit Kembhavi · Pushpa Khare

Gravitational Waves

A New Window to the Universe

 Springer

Ajit Kembhavi
Inter-University Centre
for Astronomy and Astrophysics
Pune, India

Pushpa Khare
Retired Professor of Physics
(Utkal University)
Baner, Pune, India

ISBN 978-981-15-5708-8   ISBN 978-981-15-5709-5   (eBook)
https://doi.org/10.1007/978-981-15-5709-5

This Springer imprint is published by the registered company Springer Nature Singapore Pte Ltd.
The registered company address is: 152 Beach Road, #21-01/04 Gateway East, Singapore 189721,
Singapore

Foreword

Do gravitational effects travel as waves across vast cosmic distances with the speed of light? The reason for raising this question was the expected similarity between electromagnetic theory and gravitation. In the former theory, it was well established that electric and magnetic effects travel across space at the speed of light. However, that theory was formulated differently from gravitation. The electric field and the magnetic field are two inter-related entities which travel outwards whenever an electromagnetic phenomenon occurs. For example, when alternating current is passed through a loop of wire electromagnetic waves emerge in all directions from the loop. Should we expect similar waves of gravitational nature to arise, if, say, two massive objects collide?

The question is very simply worded but not easy to answer! In the geometry-based picture of general relativity due to Albert Einstein, it is hard to separate the changes of geometry of the ambient space from the changes of gravity that arise from collision. Indeed, the question was not answered in Einstein's lifetime although there were several attempts (including claims that turned out to be wrong) to do so. Leopold Infeld, a Polish physicist who had come to work with Einstein in the Institute of Advanced Study at Princeton, narrates one incident in his autobiographical account called Evolution of a Physicist. On this occasion, Einstein found that his mathematical analysis led him to the conclusion that gravitational waves do not exist. Realising that this was an important result he agreed to give a seminar on the work. The seminar was announced but as per practice Einstein's name was withheld in the notice. [This was the usual practice since it was felt that his name would draw a horde of media persons from New York.] However, checking through the algebra of his work on the previous day, Einstein was horrified to detect a mistake of sign. When corrected for the same, his claimed conclusion did not stand. Should he cancel the seminar? After some deliberation Einstein decided to go ahead with it. At the seminar, he went through his calculations on the blackboard and concluded by showing where he had gone wrong.

In the early 1960s (Einstein died in 1955), work by several authors, including the active group at King's College, London consisting of Herman Bondi, Ivor Robinson and Felix Pirani, helped to clarify various tangled issues and led to the conclusion that gravitational waves exist. The next challenge before the general relativists was: How to detect them?

The problem here was not one of conceptual nature but practical one. Most sources were too weak to be detected by the technology available. We will briefly look at the kind of problem presented.

A comparison with the electromagnetic example will help appreciate the issue. We had mentioned an alternating current being passed through a loop of wire. A handheld loop will be able to generate electromagnetic waves detectable by a laboratory receiver. A corresponding lab source of gravitational waves may consist of two masses made to go round each other so fast that they make one revolution in a millisecond. If the separation of the system is 10 m and mass 10 kg, which are typical laboratory values, the gravitational radiation emitted by it is as low as 5×10^{-31} W.

This example demonstrates that a technology of foreseeable future will not be able to make a laboratory source for testing the existence of gravitational waves. As a short-term experiment, therefore, one needs to make the best of cosmic sources. [Recall, the law of gravitation was first tested in the cosmic setting rather than in the laboratory.] Thus, instead of two masses spinning around each other in our laboratory example, we now have two stars going around each other.

Indeed, the radio astronomers J. H. Taylor and R. A. Hulse in 1974 examined the period of the binary system made of a pulsar, with catalogue number, PSR1913+16, and a neutron star. They found that the period of the binary was decreasing at the tiny rate of 2.4 ps per second. This minute measurement was possible because the pulsar in the system provides a very accurate timepiece. The result could be explained by arguing that as the binary system emits gravitational waves it loses energy and its period decreases while the orbit size decreases. The latter effect was also found. Moreover, the rate of increase of the period agrees with the relativistic formula.

There can be other cosmic sources of gravitational waves: supernovae, for example, describe exploding stars. These will be transient since their origin is due to an exploding event of limited duration. Colliding stars, especially massive black holes, are another type on the source list. A binary made up of compact objects can be a steady source of gravitational waves whose orbit shrinks gradually. More dramatic are such binaries when they approach a limit when the two stars come close enough to merge. Additionally, we should reserve a new category of sources not yet known. For example, the quasi-steady-state cosmology which was proposed as alternative to big bang needs explosive creation events which will certainly emit gravitational waves.

In early 1960s, Joe Weber at the University of Maryland, USA had taken the lead in devising a detection system for such waves. He used metallic bars with piezoelectric strain transducers. The idea was that as a gravitational wave passes through space it carries with it information of small geometric changes. Since space geometry is influential in measuring distances of objects, the changed distances in the piezoelectric material will cause strains to arise and these can be detected and measured by electrical means. Weber had set up two such bar detectors, one in Maryland and the other in Chicago. A multiple detection system generates more confidence in a signal, for instance, if both detectors show it then the signal would be genuine, otherwise it may come from some local artefact.

Weber's system failed to detect a credible gravitational wave signal. The main reason was that his equipment was not sensitive enough. Other bar detectors like those in Rome and Perth (Australia) also failed to detect a signal. The next generation of detectors using interferometers became popular.

At the time of writing, there are laser interferometric detectors which use the Michelson interferometer for detection. In an interferometer (see Fig. 7.3), a light ray, after being split up into two parts, is made to travel along two alternative paths before combining. The interference of these rays leads to a measure of distances travelled. The passage of a gravitational wave alters the space-time geometry which the interferometer is supposed to detect. To make the equipment more sensitive, the rays are from a laser and they travel in a high-grade vacuum over large distances (the longest being 4 km).

The first and pioneering detector was built in two places in the USA, one in the northwest in Washington State and the other diagonally across the continent near New Orleans. Known as the Laser Interferometric Gravitational Wave Observatory (LIGO), the LIGO started functioning in the last decade of the twentieth century. For several years, the detector failed to detect any credible signal. The instruments were upgraded in 2015 and is now known as Advanced LIGO. There are other detectors in Europe like VIRGO and GEO, and KAGRA in Japan.

LIGO finally hit the jackpot and reported positive signals in February 2016, although the signals were first detected some months earlier on 14 September 2015 at 5.51 am. A crude calculation indicates that these arose from a collision of two black holes of 29 and 36 solar masses some 1.3 billion light years away.

This discovery opens out another major method of looking at the cosmos, perhaps as important as the use of a telescope. It is encouraging, therefore, that India is joining the LIGO network by contributing a similar detector as in advanced LIGO. An additional detector will improve our perception of the direction of the source. We should also mention that Indian expertise in data analysis played a major role in this discovery. As the hardware technology improved, the data analysis component also progressed mainly as an Indian contribution under Sanjeev Dhurandhar, the scientists trained in India by Sanjeev, and others. With LIGO-India coming on line the scope for research in this area looks promising.

With this background, it is a pleasure to welcome the book by Ajit Kembhavi and Pushpa Khare. The Marathi version of it has done well and I am sure the English version now being launched will go a long way towards making gravitational waves a household topic.

My best wishes to the book and its authors.

Pune, India   Jayant Narlikar
April 2020

Preface

In this short book, we recount the exciting and amazing story of gravitational waves. These waves were first predicted by Albert Einstein in the year 1916. The existence of the waves followed from the new theory of gravitation Einstein had formulated a year earlier, but he was not convinced of the physical reality of his prediction. Some decades had to pass before physicists were able to convince themselves that gravitational waves were indeed as real as electromagnetic waves, on which so much of modern life depends.

The effect of gravitational waves on matter is very weak and therefore extremely sensitive instruments are needed to detect them. The development of such instruments took some more decades, and finally the first detection of gravitational waves was made in September 2015 and announced in February 2016, exactly a century after Einstein made the prediction.

Our aim is to first provide the reader the background which is needed to understand the unfolding of the story from prediction to discovery, and then to describe the instruments used, the first several detections and the implications of the discoveries for astronomy and physics. We do so without using mathematics or technical discussions, laying out our arguments in a form which any discerning and interested reader should be able to understand. The story has not ended with the detection of gravitational waves. In fact, that is just the beginning of the many marvellous discoveries about the Universe to be made using the new gravitational wave astronomy. We finish the story with a description of some of the projects to be undertaken over the next few decades to that end. Our hope is that we are able to equip the reader with the knowledge and understanding to appreciate the discoveries which will be made in the future. More ambitiously, we also hope to inspire some young readers to learn enough to be able to participate directly in the development of the exciting new field.

At a time when different branches of science are making fast progress and our knowledge about the Universe is increasing in so many ways, we hope that this book will help the readers understand and appreciate the Universe through a newly opened window.

We wish to thank Prof. Jayant Narlikar for writing a Foreword for the book and Dr. Kaushal Sharma for creating many of the figures we have used. We thank Ms. Nirupama Bawadekar for obtaining photographs and other materials from various sources, along with permission for inclusion. We thank Profs. Sanjit Mitra and Tarun Sourdeep for many discussions which helped us understand better the physics of gravitational waves, and the editorial, production and other staff of Springer Nature for all the hard work that has gone into producing the book.

Pune, India   Ajit Kembhavi
                                                                                 Pushpa Khare

Contents

About the Authors

Ajit Kembhavi is Professor Emeritus and was a Raja Ramanna Fellow at the Inter-University Centre for Astronomy and Astrophysics (IUCAA), Pune. He was Director there until August 2015. He did his Ph.D. from TIFR, Mumbai, and was a postdoctoral fellow at the Institute of Astronomy, Cambridge. Professor Kembhavi is a distinguished astronomer who works on galaxies, quasars and other extra-galactic objects, various areas of high energy astrophysics including X-ray and radio pulsars. He has published a large number of research papers and several books. Professor Kembhavi has been a member of the Space Commission and is presently a member of the ISRO's Apex Science Board. He is Vice-President of the International Astronomical Union and former President, Astronomical Society of India. He was Chair of the International Virtual Observatory Alliance, Chair of the Scientific Council of the Astronomical Data Centre at Strasbourg and Chair of the Council of the Indian Institute of Astrophysics. He is a Fellow of the Indian Academy of Sciences and the National Academy of Sciences, India.

Pushpa Khare is M.Sc. (Physics), gold medalist, from Indore University. She did her Ph.D. in Astrophysics from the Tata Institute of Fundamental Research, Mumbai, and her postdoctoral work at the Max Planck Institute for Astronomy at Munich, Germany, and at the University of Manchester, UK. She then joined as Professor of Physics at the Utkal University, Bhubaneswar, Odisha, in 1983. After retiring from there in 2010, she was CSIR Emeritus Fellow at the Inter-University Centre for Astronomy and Astrophysics (IUCAA), Pune, for four years. She had been an associate of IUCAA, since its inception in 1990 till her retirement. She has mainly worked on the interstellar medium, quasar absorption lines and gravitational lensing, and has published more than 50 research papers in international peer-reviewed journals. She has been a visiting professor at several universities around the world including University of Illinois at Chicago, University of South Carolina, University of Chicago, University of Osaka, etc. She has presented her work in several national as well as international conferences including those at Shanghai, Kunming, Chicago, Columbia and Marseilles.

List of Figures

Physical Units

In the text, we have used standard abbreviations of units used to describe physical quantities such as length, time, power, magnetic field, etc. The table below provides a list of such units together with their abbreviations.

Physical quantity	Unit	Abbreviation
Length	Metre	m
Time	Second	s
Frequency	Hertz (1 cycle/s)	Hz
Angle	Arcsecond (degree/3600)	arcsec
Mass	Kilogram	kg
Energy	Erg	erg
Power	Watt	W
Temperature	Kelvin	K
Magnetic field	Gauss	G

A unit of distance used in astronomy is the *light year*. This is the distance travelled by light in 1 year at its speed of 300,000 km/s, which is 9.46 trillion km.

In the text, we have used standard symbols for multiples of units. Some multiples are listed in the table below.

Multiple	Name	Abbreviation
Mega	Million 10^6	M
Kilo	Thousand 10^3	k
Deci	Ten	d
Centi	Hundredth 10^{-2}	c
Milli	Thousandth 10^{-3}	m
Micro	Millionth 10^{-6}	μ

(continued)

(continued)

Multiple	Name	Abbreviation
Nano	Billionth 10^{-9}	n
Pico	Trillionth 10^{-12}	p
Femto	Quadrillionth 10^{-15}	f

In this notation, for example, a thousand seconds would be 1 ks, 1 billionth of a second would be 1 ns and 1 trillionth of a metre would be 1 pm.

Chapter 1
Introduction

Abstract On 11 February 2016, a momentous announcement of the first detection of gravitational waves was made by David Reitze, the Director of the LIGO Laboratory. The existence of these waves was first predicted by Albert Einstein a century earlier in 1916. But these waves are extremely weak and had remained undetected in spite of several efforts. They were finally detected by the Advanced LIGO detectors located in Hanford, Washington State and Livingston, Louisiana in the USA on 15 September 2015. The announcement of detection was made after long and careful analysis of the data to establish that the weak signal was indeed due to a gravitational wave and not due to some spurious noise in the system or seismic disturbance. Astronomers all over the world were greatly excited by the discovery, which was a triumph of the extreme technical capability scientists established through decades of hard work. The discovery verified Einstein's century-old prediction and enabled for the first time the direct detection of a black hole and a black hole binary system. It also opened the new window of gravitational wave astronomy to the Universe, which until then could be observed only through electromagnetic waves, cosmic rays and neutrinos coming to us from distant objects. In the first chapter, we summarise the announcement of the discovery and set the stage for the rest of the book.

On the 11 February 2016, scientists and journalists from all over the world were gathered for a press conference called by the LIGO Labs (LIGO: Laser Interferometric Gravitational Wave Observatory) in the USA. Under the LIGO Labs there are two gravitational wave detectors which for years had been trying to detect the waves. The detectors had gone through a period of upgradation, and had been operating with much enhanced sensitivity since September 2015, so it was widely believed that there would be an announcement of a detection at the press conference. A small number of scientists, journalists and media persons were fortunate enough to be present at the venue of the announcement in Washington D. C. But thousands more were in remote attendance in small and large groups all over the world, as the event was broadcast over the Internet. In all such gatherings, the excitement was palpable.

The event in Washington D. C. opened with some remarks by the Head of the National Science Foundation, USA, which funds the LIGO Labs. In a few minutes,

David Reitze, the Director of LIGO Labs, was handed the microphone and he made the much awaited announcement: 'Ladies and gentleman, we have detected gravitational waves, we DID IT!'. This was met with thunderous applause not only at the venue but at all places in the world where the event was being watched. The audience in the several packed auditoria in the world stood up clapping and yelling. Everyone present there was thrilled and felt extremely privileged to have been able to listen to the announcement as it was made. Some of the people in the audience had made seminal contributions to the theory of gravitational waves and the data analysis required to detect the signal. Many others present had made important contributions to the building of the instruments, observations and analysis and were authors of the research paper in which the discovery was announced.

Why was everyone so excited? What are gravitational waves and why is their detection so important? The purpose of the present book is to answer these questions in some detail, but in a manner which would be accessible to everyone who may be interested, regardless of their background. While we will develop the story and describe all terms and concepts and ideas over several chapters of the book, we will present a short introduction here for the reader. The reader may be unfamiliar with some of the terms and concepts appearing in this chapter. All these will be described in the following chapters.

Gravitational waves were first predicted by Albert Einstein in 1916, from his theory of gravitation, which he had completed a year earlier in 1915. Strong gravitational waves are produced by certain cosmic objects. They propagate with the speed of light as ripples in the structure of space-time and lead to changes in the dimension of an object in their path. The effect produced is extremely tiny and is very difficult to detect. It was realised in the 1960s and 1970s that gravitational waves could be detected by laser interferometers. Construction of the first such devices started in the mid-1990s in Hanford in Washington State and Livingstone in Louisiana. The LIGO interferometers began gravitational wave searches in 2002, which continued for about a decade. While no detection was made in this period, it was possible to develop deep understanding about the possible nature of gravitational wave sources and the improvements needed in the instruments to successfully detect gravitational waves.

A major upgradation of the detectors which would make possible observations at higher sensitivities was undertaken in 2008 and the advanced LIGO detectors were completed in 2014. Observations with these new detectors were to begin from the middle of September 2015. But during the engineering run a few days before the planned commencement of observations, a signal was observed which could be a gravitational wave detection. Since such signals are extremely weak, very careful analysis of the data is required to establish that the signal is indeed produced by a gravitational wave and not due to any spurious effect. After this task was completed and the result was accepted for publication in a reputed scientific journal called Physical Review Letters, the announcement was made at the press conference mentioned above on 11 February 2016. Why is the detection so important, and what new insights does it provide us that the scientific community, and the public at large, was so excited?

The Italian physicist Galileo Galilei was the first person to use a telescope to observe the sky. He made astonishing discoveries like sunspots, the Moons of Jupiter, the phases of Venus, and stars in the Milky Way. These were totally unexpected phenomena which revealed objects in the sky that had remained unobserved until then. His observations wholly changed the course of science. In the centuries after Galileo, bigger optical telescopes and better instruments to be used with them were built, which greatly increased our knowledge of the Universe.

In the second half of the twentieth century, technological developments made it possible to build radio, millimetre, infrared, ultraviolet, X-ray and Gamma ray telescopes, some of which have to be used from space. These telescopes all detect electromagnetic waves and together provide us with a multi-wavelength view of the Universe and increasingly deep understanding of how it works. The LIGO detectors work very differently: they detect gravitational waves which are ripples in the structure of space-time, rather than being electromagnetic in nature. So the information that we get from detectors like LIGO simply cannot be obtained using any of the other telescopes mentioned. This is evident from the nature of the first source of gravitational waves detected. It was a system consisting of two massive black holes in orbit around each other, which in a fraction of a second spiralled inwards and merged to form a single black hole, in the process emitting the signal that was detected on 14 September 2015. Since black holes do not emit any electromagnetic radiation, none of the other telescopes could detect the merger, and it would have gone undetected if LIGO was not in operation.

Metaphorically speaking, it was as if so far we had only one window through which we could 'see' the Universe, using the electromagnetic radiation that cosmic objects emit or absorb. However, there are events occurring in the Universe, like the binary black hole merger, which may not emit electromagnetic waves and so would be invisible to our eyes and to our conventional telescopes: they simply cannot be 'seen' through the only 'window' which was open to us. Like Galileo's small telescope did, gravitational wave detectors provide us wholly new views of the Universe, taking us from multi-wavelength astronomy to what has become known as multi-messenger astronomy. So we can say that a second 'window' in the form of gravitational waves has opened for us through which we can see different views of the Universe.

The first gravitational wave detection proved that Einstein's prediction of the existence of gravitational waves, made in 1916 a century before the detection, was correct. It was the first direct detection of a black hole and the first detection of a binary black hole. The masses of the two black holes in the binary are much larger than the masses of the black holes inferred to exist in X-ray binaries. Subsequent to the first detection, many other detections have been made until the present. It appears that there is a significant population of black holes and binary black holes with such masses, which are greater than the black hole masses known earlier, but much smaller than the supermassive black holes known to exist in the centre of galaxies. It is clear that a wholly new chapter has begun in astronomy. The detailed form of the signals has yet again confirmed the correctness of Einstein's theory of gravitation, which is known as the general theory of relativity and is considered to be one of the most beautiful creations of the human mind. The very first detections have already provided us with so much new knowledge and the possibilities for the future are endless, which explains the excitement generated by the discovery.

This book is primarily about gravitational waves, their emission and detection. The understanding of these topics requires some background in the theory of electromagnetism, particularly electromagnetic waves, and in Newton's theory of gravitation. We discuss these topics in early chapters of the book. Gravitational waves are a prediction of Albert Einstein's general theory of relativity, which is a theory of gravitation quite different from Newton's theory. Understanding general relativity requires some background in Einstein's special theory of relativity. We discuss all these topics briefly, focussing on those areas which are most relevant to the main topic of the book. We then describe the theory of gravitational waves, their emission, various types of gravitational wave sources and the binary pulsar which provided the first evidence that gravitational waves are emitted by compact binaries as per the prediction of Einstein's theory. In the final chapters of the book, we discuss gravitational wave detectors, the first detection of gravitational waves, a few other sources that have been detected and future gravitational wave detectors.

The topics described in the book all require advanced mathematics and physics for their full understanding. Our aim in writing this book is not to provide a technical treatment with mathematical development, but to introduce the reader to the recent excitement in the field of gravitational waves in a simple and hopefully lucid manner. We use only qualitative arguments, and with the help of analogies and diagrams try to convey some profound ideas which have been developed over centuries, and their rapid realisation over the last few decades. It is amazing that a theory which was fully developed a century ago is at the heart of some of the most exciting advances of our time, which is technologically so different from the early twentieth century. We hope to convey some of that excitement to anyone who has the patience to read our book.

Chapter 2
Electromagnetic Radiation: The Key to Understanding the Universe

Abstract In this chapter, we describe electric and magnetic fields, their unification by Maxwell, electromagnetic waves, Albert Einstein's special theory of relativity and how we get information about cosmic objects through electromagnetic waves emitted by them. The work of Maxwell provided the theoretical framework for understanding electric and magnetic fields which had been experimentally studied for centuries. He unified the two fields into the electromagnetic field and predicted the existence of electromagnetic waves, which are the basis of much of modern technology. A fuller understanding of Maxwell's theory is obtained through the special theory of relativity. Depending on its wavelength, the radiation is known as Gamma rays, X-rays, ultraviolet, visible, infrared, microwave or radio radiation. Stars emit mainly ultraviolet, visible and infrared radiation. Through a detailed analysis of the electromagnetic radiation coming from astronomical sources, we can determine their properties like mass, temperature, distance, size, chemical composition, radial velocity, age and magnetic field. Almost all our knowledge of the Universe obtained till 2016 came through the study of electromagnetic radiation reaching us from astronomical sources.

2.1 Introduction

In all our life, we unknowingly depend on one natural phenomenon more than any other and that is electromagnetism. This encompasses our entire existence. First of all, out of our five senses, the one which arguably is the most important and without which human beings could never have reached the stage of development that they are at is seeing. We see because our eyes are sensitive to light, which is an electromagnetic wave. Fortunately for us, the stars, including the one most important for our very existence, the Sun, emit these waves. So all our early information about the world around us and the heavens above us came through seeing objects with the help of these waves. Next, all of the chemical reactions that take place inside our bodies, all the signals that the brain gives to our body parts, essentially all our body functions make use of electromagnetism in one way or another.

A. Kembhavi and P. Khare, *Gravitational Waves*,
https://doi.org/10.1007/978-981-15-5709-5_2

All the modern devices that we use these days for communication and travel such as the phone, the car, the train, the aeroplane; for work such as the computer, the laptop, the tablet, or machinery of most kinds; for entertainment such as the television, the radio, the video games; for cooking such as electric or induction cookers and utility equipment such as washing machines, refrigerators, light bulbs, tubes, fans, air conditioners; machines in our factories, instruments in our hospitals all depend on the electromagnetic phenomenon.

How and when did we discover electromagnetism? The first reference to lodestones, which are natural magnets, is found in the works of the Greek philosophers of the sixth century BC. Similarly, the ancient Greeks also knew of amber, which develops electric charge when rubbed; we all know that a comb after use attracts paper bits, which is also due to the charge developed on the comb. However, the electromagnetic phenomenon was theoretically understood much later, mostly in the nineteenth century, with the efforts, including laboratory experiments, of a number of scientists. Obviously, we have made tremendous progress in understanding this phenomenon and in making use of it to develop gadgets which have made our life not only comfortable but greatly enriched. Most of this progress has been made in the twentieth century.

Here, we are only concerned with a particular aspect of electromagnetism, namely, electromagnetic waves, which has been the only tool for us to observe and understand the Universe until 2015, that is before the detection of the *gravitational waves*. One important fact to note here is that as far as understanding the Universe is concerned, the only option available to us is passive observation. We cannot perform experiments on parts of the Universe as we do in our laboratories. The reason is that the objects that we are concerned with in astronomy, like planets, stars and galaxies, are all very far away and are mostly unreachable. We have indeed managed to land on a few astronomical objects in the Solar system, like the Moon and asteroids, and have managed to bring material back to the Earth from the Moon, a comet and an asteroid. But barring these exceptions, it is only by studying the radiation, i.e. the electromagnetic waves coming from the heavenly objects that we have to gain information about the objects emitting or absorbing the radiation.

There is another source of information, namely, the high-energy particles called cosmic rays coming from space, but these are more difficult to interpret. This is because most of these particles are electrically charged and their path gets affected by the magnetic fields that they come across during their travel to the Earth, making it impossible to locate their source. We also receive some particles which have extremely small mass and no electric charge, called neutrinos, from astronomical sources. We have been able to detect neutrinos coming from the Sun, and in 1987, from a supernova explosion. However, most if not all of our information obtained so far about the Universe is through the study of electromagnetic waves reaching us. Human beings are highly ingenious and have found ways to utilise this radiation maximally. We can actually measure the mass, temperature, velocity, chemical composition and other properties of stars, galaxies and other cosmic objects, as well as the distances to many of these by the analysing this radiation.

2.2 Electromagnetism

Important contributions towards understanding electromagnetism were made by scientists like Charles-Augustine de Coulomb, Hans Christian Øersted, Karl Friedrich Gauss, Michael Faraday, André-Marie Ampère and others starting from the last few decades of the eighteenth century. The fundamental quantities which describe this phenomenon are electric charges and currents, and electric and magnetic fields which we describe below.

2.2.1 Electric Charges, Currents and Magnets

There are two kinds of electric charges, which are conventionally called positive and negative. As was noticed long ago, a positive charge can be produced on a glass rod by rubbing it with silk, while negative charge can be produced on amber by rubbing it with fur. Of course, the charges are not really created, they are merely separated. The glass rod accumulates positive charge while an equal amount of negative charge is accumulated on the silk. Similar separation occurs on amber and fur. A very basic property of the charges is that two bodies which have the same kind of charge (both positive or both negative) have a repulsive force acting between them, while bodies which have different kinds of charges attract each other. This repulsive or attractive *electrostatic force* acts between bodies even when they are in vacuum, i.e. the force does not need any medium to act through. The strength of the force of attraction or repulsion between two electrically charged bodies was first described by Coulomb and is known as Coulomb's law. According to this law, the force is proportional to the magnitude of each charge and diminishes with the square of the distance between them. The meaning is that (i) if the magnitude of one charge is doubled, the force gets doubled and (ii) if the distance between the charges is doubled, the force decreases to a quarter of its original value; if the distance is tripled, the force diminishes to a ninth of its values and so on. This is called the inverse square law.

It became known during the first two decades of the twentieth century that all matter is ultimately made up of atoms, which themselves consist of three kinds of elementary particles: protons which have a positive electric charge, electrons which have a negative electric charge and neutrons which have no charge and are said to be neutral. The charges on the proton and the electron are exactly the same in magnitude. A body which is positively charged has an excess of protons and a body with negative charge has an excess of electrons. When a positively charged body is brought in contact with an equally negatively charged one, the charges can neutralise each other. Rubbing a glass rod with fur separates the electrons from some of its atoms which accumulate on the fur.

We are all familiar with magnets and frequently use them to fix small objects to refrigerator doors and other metallic surfaces. Every magnet has two *poles*, one of which is conventionally called the north pole of the magnet and the other one the south pole. Unlike positive and negative electric charges, the poles cannot exist separately, and every magnet has both the poles. If a magnet is broken into two pieces, each piece has both poles. Similar to electric charges, like poles repel while unlike poles attract each other.

Coulomb
Charles Augustin Coulomb (1736–1806) was born in France in an aristocratic family. He studied mathematics in a college in Paris. After finishing his education he joined the French army as an engineer. For 20 years after that he was stationed in several different places. His experiments on friction and stiffness of ropes resulted in the publication of his major work 'Theory of Simple Machines' in 1781. He was awarded the Grand Prix of the Acadmie of Sciences. After that he worked on physics rather than engineering projects. Between 1785 and 1791 he published important papers on electricity and magnetism. The unit of electric charge has been named as Coulomb in his honour. His name is in the list of 72 French scientists, mathematicians and engineers engraved on the Eiffel Tower.

Travellers have been using magnetic compass to know directions since ages. It is a thin, needle-like magnet which is mounted in such a way that it can rotate freely about a vertical axis. The north pole of the magnetic needle points approximately to the geographic north pole of the Earth and the south pole of the magnet points approximately to the geographic south pole of the Earth. The reason is that the Earth itself acts like a bar magnet, which is aligned approximately in the south-north direction.

We are all familiar with the flow of electric current in a wire. The current is due to the motion of the negatively charged electrons in the wire. The current carries energy, which is why it is able to heat the tungsten filament in a light bulb, making it glow. But the property of an electric current which is most relevant to us here is that it produces magnetic effects. If a magnetic compass or some other magnet is brought close to a current-carrying wire, then the magnet is seen to feel a force, as if another magnet were present in place of the current-carrying wire. The strength of the magnetic effects produced by a current was first described by Ampère and is known as Ampère's law. If there are two wires, each carrying a current, then a force is felt between them. This can be understood to be due to the magnetic effects produced by each wire. Another phenomenon is observed when a magnet is moved back and forth through a loop of wire. It is found that the motion of the magnet generates a current in the loop, as if the loop were connected to a battery. This effect was studied by Faraday and was described in the form of an equation known as Faraday's law. Ampère's law is at the basis of electric motors, while Faraday's law is the basis for the generation of electrical power using turbines.

A question which could be asked here is: are the magnetic effects generated by currents different from the effects of bar magnets, say? When the supply of electricity to a wire is stopped, the magnetic effects associated with the current vanish. On the other and, the effects produced by a bar magnet appear to be permanent, and they cannot be switched off. The connection between the two was understood only after the discovery of atomic physics and quantum mechanics. It turns out that the magnetic fields in permanent magnets are generated by tiny electrical currents which are caused by the motion of electrons around the atomic nucleus and other related phenomena. So atoms of certain elements, depending on the number of electrons in them, can act as tiny magnets. But in most materials these tiny magnets are randomly oriented and their effects get cancelled and the substance as a whole does not posses magnetic properties. It is only in some substances that these tiny magnets can align themselves so that the substance as a whole has magnetic properties. The natural magnets are made of these substances. In materials like iron, the tiny magnets can be aligned temporarily by using another magnet or using a current.

When the charges and natural magnets were first discovered and studied, it was not realised that the two phenomena were related to each other. From the experiments conducted in the eighteenth and nineteenth centuries, it was realised that moving charges generate magnetic effects, while a moving magnet produces currents. Electricity and magnetism were, therefore, related to each other through Ampere's and Faraday's laws. How deep this connection is became clear only after the work on electromagnetism by James Clarke Maxwell and later the development of the special theory of relativity by Albert Einstein, which are described below.

Faraday

Michael Faraday (1791–1867) was born in a small town in England. His father was a blacksmith and the family was very poor and faced starvation at times. Michael had to leave school at the age of 13 because of poverty. At the age of 14, he got a job with a book binder and read a number of books. During that time he heard lectures by Sir Humphrey Davies. His meticulous notes taken during the lectures impressed Sir Humphrey who accepted him as an apprentice. After that Michael did not look back and continued to perform experiments one after another. He started his career as a chemist and discovered several new organic compounds including Benzene. He also invented an early form of the Bunsen burner which is now extensively used in laboratories all over the world. However, his most important work was in the field of electromagnetism. He was awarded a Doctorate by Oxford University in 1832. The unit of capacitance is named as Faraday in his honour.

Ampère

André-Marie Ampère (1775–1836) was born in a well-to-do family in France. His father did not send him to school but instead educated him at home. He built a good library at home for the purpose. According to Ampère, he had got all the necessary knowledge of physics and mathematics by the age of 18. He did fundamental work on current electricity, remaining in the teaching profession throughout his life. He held prestigious positions at well-known institutes in Paris. The unit of current is named Ampere in his honour. His name is among the names of 72 French scientists, mathematicians and engineers engraved on the Eiffel Tower.

2.2.2 The Electromagnetic Field

We have introduced above the concepts of electric charges, currents and magnets. We would now like to understand the concepts of electric and magnetic fields, which are needed for the full understanding of electromagnetism. What are these fields? A field describes the effect that a charge or current or magnet has in the surrounding region. These fields have their own reality, separate from the sources which generate them and propagate electromagnetic effects over great distance even through vacuum.

We can say that an electric charge produces an electric field. If a charged particle, which we will call a *test particle*, is kept in an electric field, it will experience a force of attraction or repulsion, depending on the signs of the source charge and the charge on the test particle. The strength of the electric field at any point due to a single electric charge is determined by the force experienced by the test particle kept at that point and so is given by Coulomb's law which we have described above. The larger the magnitude of the source, i.e. the charge which produces the field, the greater is the strength of the field. Also, as the distance from the source increases, the strength of the electric field decreases as per the inverse square, as stated by Coulomb's law. The direction of the force on a positively charged test particle present at a point in space is said to be the direction of the electric field at that point. When there are many charges present, the total electric field can be calculated as the sum of the electric fields produced by each charge.

Oersted

Hans Christian Oersted (1777–1851) was born in Belgium. Even as a child he developed interest in science while helping in his father's pharmaceutical laboratory. He was home schooled and directly joined the university for higher studies. The relation between electric and magnetic phenomena was noticed by him accidentally. While he was performing some experiment with electric current in his laboratory in 1821, a magnetic needle happened to be nearby. Oersted noticed that the needle changed its direction when the current was switched on or switched off. With further studies, he discovered the principles of electromagnetism. He also made significant contributions to chemistry. He produced aluminium for the first time. The unit of magnetic field has been named the Oersted in his honour. He was also a writer and poet and published a book of essays and poems.

Similarly, a magnet can be considered to produce a magnetic field, which affects other magnets present in the vicinity. In addition, the magnetic field also exerts a force on an electric charge placed in it, if the charge is moving. Electric and magnetic fields taken together are known as the electromagnetic field. The two kinds of fields are really not distinct, as we have seen from Ampere's law and Faraday's law. More generally, it is observed that a changing magnetic field produces an electric field and a changing electric field produces a magnetic field. How much force a test particle will experience in an electric or magnetic field will, of course, depend on the strength of the fields, but it will also depend on the test particle's charge, position and velocity. A particle which is at rest feels no force due to a magnetic field. Neutral particles, i.e. particles with zero charge, do not experience any force in an electromagnetic field.

While the fields we have described above seemed to be anchored in charge, currents and magnets, as already mentioned, they have their own reality. We will see below that electromagnetic fields can propagate through space as waves travelling at the speed of light, carrying energy. Once produced the fields continue to propagate and carry energy, even if the sources which produced them no longer exist.

2.2.3 Maxwell's Equations

By the second half of the nineteenth century, it had become known that there are four fundamental laws which describe the relation between the electric and magnetic fields and their sources, i.e. the distribution of charges and currents which produce the fields. These laws are mathematically represented by equations which are known after the scientists who wrote them first as described above: Gauss's law for electric fields (which is essentially Coulomb's law expressed differently), Gauss's law for magnetism, Faraday's law and Ampere's law. These equations were studied in detail by the great Scottish physicist James Clarke Maxwell. He found that the equations, when taken together, were not mathematically consistent in circumstances where the sources of the fields change with time. Maxwell found that he could restore mathematical consistency by adding an extra term, known as the displacement current, to the equations describing Ampere's law. This appears to be a small change, particularly in modern mathematical notation, but it has the most profound consequences. The new term in effect says that changing electric and magnetic fields can themselves act as sources of the fields.

Gauss

Carl Friederich Gauss (1777–1855) was born in a very poor family in Germany. His parents were illiterates. However, his intelligence became known at a very young age and attracted the attention of the Duke of Brunswick who took the responsibility of his education. At the age of 3, he used to work out addition and subtraction orally. He used to surprise his teachers by solving problems given to the class very fast using novel methods. He started discovering new formulae in mathematics at the age 10. At the age of 24, he wrote a book on mathematics. This book is considered to be one of the best books on the subject and Gauss is considered to be the greatest mathematician of all time. He also made significant contributions to physics and astronomy. After his death, his brain has been preserved in a laboratory in Groningen, Netherlands. The unit of magnetic induction has been named as Gauss in his honour.

The new term is responsible for the all important phenomena of electromagnetic waves, which would have remained unexplained without the addition of this term. Maxwell first introduced the term in 1861, and in 1865 he predicted the existence of electromagnetic waves. The mathematical notation at that time was cumbersome, and he had to work with 20 equations in all. It was some years later that Heaviside and others used new mathematical notation for writing the equations. They were then reduced to four equations in the simple and elegant form that we use today and know as *Maxwell's equations* of electromagnetism, because of the all important terms added by Maxwell in one of the equations.

We have described above how electric and magnetic fields exert a force on electric charges. The equation which describes this force is known as the Lorentz force law. Maxwell's four equations and the Lorentz law together are the mainstay of electromagnetism and allow us to understand and express in exact mathematical form all electromagnetic phenomena. Changes to this formalism are required only for understanding atomic and molecular physics and elementary particles.

A consequence of Maxwell's equations is that an electric charge which is accelerated emits *electromagnetic radiation*. If the charge is at rest or moves with an unchanging velocity, then it does not radiate. All electromagnetic radiation that we observe can be traced ultimately to accelerated charges or magnets. The radiation can be emitted by just one charge, a group of charges, a changing electric current, an appropriately rotating magnet or more complex sources. The fields emitted in this manner travel through space as waves, which we describe below. The fields carry energy and continue to propagate unless they are absorbed by matter. The fields can exist even if the sources that produced them cease to exist.

Maxwell
James Clark Maxwell (1831–1879) was born in a well-to-do family in Scotland. His curious and sharp mind was evident from a young age. At the age of 14, he wrote his first research paper. He had his education first at the University of Edinburgh and then at the University of Cambridge. He was appointed as a professor at the age of 25 in Marischal College in Aberdeen Scotland and later in King's College, London. In 1871, Maxwell was appointed as Professor of Experimental Philosophy in Cambridge, where he set up the Cavendish Laboratory, which developed into one of the great research centres in the world. He carried out research work at the highest level in several subjects including electromagnetism, thermodynamics and statistical mechanics. His work on electromagnetism, in which he unified magnetism, electricity and optics is considered important among all his work and for physics as a whole. He also made significant contributions to other fields including astronomy and colour photography. He mathematically proved that the rings around Saturn are made of dust particles and produced the first ever colour photograph. Einstein has said that the world had changed forever due to Maxwell's work on electromagnetism.

2.3   Waves

We know several different kinds of waves. Easiest to see around are the water waves, the ones which are produced when we throw a stone in a water body. These travel outwards from the point the stone falls in to the water. At any given point, the water particles go up and down with great regularity. Then there are waves on a string or a rope which are produced when we pluck or shake it. Particles of the string shift from their normal positions with great regularity. Similarly, we have sound waves in which again particles of the medium, like air, oscillate about their undisturbed positions. While no single particle moves far from its rest position, as time passes the wave itself propagates further and further away from the source of the sound.

A wave of any type is characterised by a property called the wavelength. It is the distance between two consecutive particles in same state of motion, for example, the distance between particles of water in case of water waves and of the string, in the case of waves on a string, having maximum or minimum displacements along the same direction. This is shown in Fig. 2.1. The maximum or minimum displacement of the particle from its rest position is known as the amplitude of the wave. Waves can have different wavelengths as shown in the figure. Another property of a wave is its frequency which is the number of wavelengths that crosses a point in space in unit time. The product of the wavelength and frequency is equal to the speed of the wave in the medium, which is the total distance travelled by the wave in unit time.

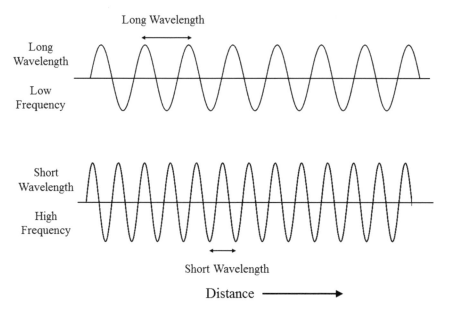

Fig. 2.1 Waves with different wavelengths/frequencies. The figure shows the magnitude of some oscillating entity at a given time, for example, the displacement of water particles from their undisturbed positions, as a function of distance from some starting point. A figure showing the variation of the entity with time, at a given space point, will look exactly the same. Image credit: Kaushal Sharma

The speed of a wave depends on the properties of the medium that it travels through, like water or the string through which the waves travel. For a given medium, the speed of a wave is fixed, and so, the higher the frequency, the shorter is the wavelength, and vice versa. Sound waves, for example, have a speed of about 1000 kilometres per hour (km/h), which is somewhat more than the speed with which passenger jetliners travel. The speed depends on the conditions of the air through which the sound travels, like its temperature and humidity.

Mathematically, all types of waves can be described by a particular type of equation which is known as the *wave equation*. When we describe waves on a string, we have a one-dimensional wave equation, because a string is an object with just one dimension, its length. On the surface of water we have two-dimensional waves. On the other hand, when sound waves propagate in all directions in space from their source, we need a three-dimensional wave equation. The wave equation tells us how much the displacement of a particle of the medium is at any time, in a given place in the medium, and how the disturbance propagates outwards from the source which created it.

2.3.1 Electromagnetic Waves

After Maxwell developed the equations describing the electromagnetic field, he studied them in various ways, and in the process he made a startling discovery. He found that he could combine these equations to form a wave equation. In the wave equations we considered above, it was the displacement of particles which was described by the equation. In the case of electromagnetism, Maxwell found that it is the electric and the magnetic fields that change periodically with time and distance, and are described by the wave equation in three dimensions. So Maxwell predicted the existence of new types of waves.

It can be shown by solving Maxwell's equations that the directions of electric and magnetic fields in a wave are perpendicular to each other and both of these are perpendicular to the direction of propagation of the wave. Such waves are called transverse waves. This is shown in Fig. 2.2. The electric and magnetic fields are in phase, i.e. both happen to be zero or maximum at the same instant of time and at the same point in space. The magnitudes of the fields in an electromagnetic wave are extremely small as compared to the values of the fields encountered in various electrical devices that we use. The beauty of electromagnetic waves is that once generated, the systematic oscillations of the fields can continue indefinitely even if the sources that created them disappear. This is because of the fact that a changing electric field generates a magnetic field and vice versa. The waves, therefore, have their own existence. When a wave encounters some obstacle in its path it may get absorbed in the obstacle and no longer exist, or it may be scattered, i.e. change its direction and continue its motion along the new direction. These waves carry energy and are also known as electromagnetic radiation. The electromagnetic waves display various phenomena with which we are familiar from the study of light as a wave.

All other waves known to Maxwell required a medium to travel through, like water waves require water and sound waves require air or any other material. Maxwell, therefore, assumed the presence of a medium, called the *ether*, for the propagation of electromagnetic waves. This medium was believed to be present everywhere in space including in vacuum, which is devoid of any other known material. This ether, of course, is not related to the chemical ether with which we are all familiar. From his equations, Maxwell was able to predict the value of the speed of the electromagnetic waves in ether, which turned out to be very close to the speed of light, which had been measured fairly accurately by then to be about 300,000 kilometres per second (km/s). It had also been established through numerous experiments that light is a wave. So Maxwell speculated that light is an electromagnetic wave. However, it was not immediately clear if the electromagnetic waves themselves were real and could be identified with light. It was only in 1887, 9 years after Maxwell's death, that Heinrich Hertz could detect electromagnetic waves of a form which we now know as radio waves. He also measured their speed which was found to be equal to the speed of light. All properties of electromagnetic waves were experimentally shown to be as per the predictions of Maxwell's theory. So the phenomena of electricity, magnetism and light were all unified through Maxwell's equations, which was a great achievement for theoretical physics, and for Maxwell.

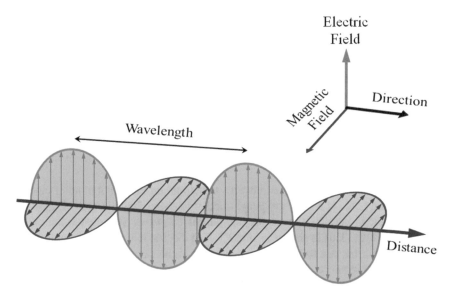

Fig. 2.2 Electric and magnetic fields in an electromagnetic wave. The electric field is shown in red while the magnetic field is shown in blue. The vertical and horizontal arrows show the magnitude and direction of electric and magnetic fields, respectively. The black arrow shows the direction in which the wave is travelling. Image credit: Redrawn by Kaushal Sharma based on a figure in Wikipedia under a Creative Commons Licence (https://en.wikipedia.org/wiki/File:EM_spectrumrevised.png)

The theory described above is known as a classical theory, in which light travels as waves. There exists a branch of physics, called *quantum mechanics*, which is applicable to systems of atomic dimensions or smaller, for which classical theories fail. As per this theory, there is a wave-particle duality and light can be considered either as an electromagnetic wave or as a collection of particles which are named *photons*. Some of the experiments performed in the laboratory like *interference* and diffraction are consistent with light being a wave while some others like the photoelectric effect, need it to be made up of particles. According to the quantum theory, the energy of a photon is given by the product of the frequency of the equivalent wave and a constant called *Planck's constant*, which plays a very important role in quantum mechanics. This means that waves with higher frequency or smaller wavelength have higher energy. This can be seen in a day-to-day experience. It requires more energy to produce higher frequency waves on a rope as compared to lower frequency waves.

It is worth mentioning that Albert Einstein was awarded the Nobel prize for his explanation of the photoelectric effect which uses the particle nature of light, and not for his path-breaking work on relativity which we will be describing below. The Nobel committee thought relativity was not proven sufficiently well and also they failed to recognise its importance. However, the importance of relativity cannot be undermined. It is only because of his relativity theory that we can fathom our Universe at a deeper level.

Coming back to the electromagnetic waves called light, these are the waves that our eyes are sensitive to and which enable us to see the splendour of the heavens. Serious astronomical observations using naked eyes have been going on since several millennia. These had enabled several civilisations more than 2000 years ago to notice the difference between planets and stars and the regularity of the occurrences of eclipses and to even predict them. However, serious quest towards 'understanding the Universe' started only after Galileo first pointed his telescope towards the sky about 400 years ago. Since then, mankind has not looked back. Bigger and bigger telescopes have been constructed to look deeper and deeper into space and earlier in time.

2.3.2 Types of Electromagnetic Waves

The electromagnetic radiation that our eyes are sensitive to is called light or the *visible radiation*. These have wavelengths roughly between 400 and 700 nanometres (nm); one nanometre is 10^{-9}th of a metre. We perceive waves of different wavelengths in this range as having the different colours displayed in a rainbow, i.e. VIBGYOR, starting from violet and continuing through indigo, blue, green, yellow, orange to red. The violet colour has the shortest wavelengths, around 400 nm while the red colour has the longest wavelengths around 700 nm. However, electromagnetic waves come in an infinite range of wavelengths. These can be smaller than a nanometre and can be larger than a kilometre. Electromagnetic radiation has been given different names, each covering a range of wavelengths. Thus, as seen in Fig. 2.3, waves starting with smallest wavelength (highest energy) to largest wavelengths (least energy) are called *Gamma rays, X-rays, ultraviolet, visible, infrared, microwave* and *radio waves* in that order.

The radiation that we receive from the Sun is a mixture of electromagnetic waves of a large number of wavelengths. We can separate or spread out visible wavelengths by passing sunlight through a glass prism. The resulting picture is similar to the one at the bottom of Fig. 2.4 and is called the spectrum of the Sun. This was first obtained by Sir Isaac Newton. The rainbow is also a *spectrum* of the Sun, produced naturally. Here the rain drops act as tiny prisms and collectively produce the spectrum. The rainbow and the laboratory spectrum shown at the bottom of Fig. 2.4 are images of the spectrum. These are low-resolution spectra of the Sun. Using more sophisticated equipment in the laboratory, one can obtain a high-resolution spectrum. A part of such a spectrum of the Sun covering visible wavelengths is shown at the bottom of Fig. 2.4. This shows some dark lines which are called *absorption lines* which yield important information about the Sun as we will discuss below. The spectrum can be displayed in the form of a graph of energy received by us as a function of wavelength. The upper portion of Fig. 2.4 shows such a graphical spectrum of the Sun. The energy received is maximum at about 450 nm. Note that the energy emitted is relatively small

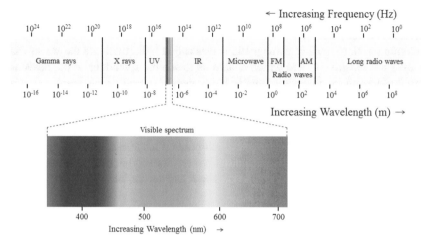

Fig. 2.3 The full range of wavelengths, in metres, for electromagnetic radiation. The ranges of different types of waves are indicated. The inset shows the visible radiation. Here wavelengths are in nanometres (nm). Image credit: Redrawn by Kaushal Sharma based on a diagram by Philip Ronan, Gringer in Wikipedia under a Creative Commons License (see https://en.wikipedia.org/wiki/File:EM_spectrumrevised.png)

at certain wavelengths, the spectrum showing a dip at these wavelengths. These are the absorption lines which are seen at the same wavelengths in the spectral image at the bottom.

Different astronomical sources emit different types of waves in different proportions. Even though we can actually 'see' only a very tiny fraction of the full range, namely, the visible radiation, all other types of waves also carry important information about the sources. Often the information is complimentary to that obtained through the study of visible light. So astronomers use different types of telescopes and detectors to gather and study these radiations. Most of these waves, with the exception of visible and near-infrared radiation and the radio waves, get absorbed while passing through Earth's atmosphere and so the telescopes used for their study have to be placed above Earth's atmosphere aboard artificial satellites.

2.3.3 Thermal Emission

Absolute Temperature
For everyday purposes like measuring ambient temperature or body temperature, we use either the Centigrade (Celcius) or the Fahrenheit scale of temperature. For scientific purposes, a different scale known as the absolute or the Kelvin scale of temperature, named after Lord Kelvin (Sir William Thomson), was used. He was a nineteenth-century physicist and engineer who made

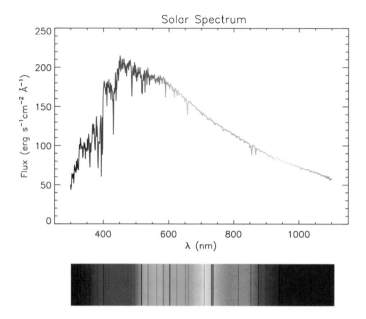

Fig. 2.4 Image and graphical representation of the Solar spectrum. Image credit: Redrawn by Kaushal Sharma based on diagram in Wikipedia under a Creative Commons License (see https://commons.wikimedia.org/wiki/File:Fraunhofer_lines.svg)

important contributions to thermodynamics. Absolute temperature has a simple relation to the temperature in Centigrade: the absolute temperature is obtained by adding 273.15 to the corresponding temperature in Centigrade. So the freezing point of water, which is 0° Centigrade, has absolute temperature of 273.15 K, and the boiling point of water on the absolute scale is 373.15 K. Absolute temperature is not just another temperature scale, it has deep significance and all physical laws which involve temperature use the absolute scale. The zero of this scale has special significance. In classical physics, all molecular motions cease at this temperature, which is a state of zero energy. It follows from the third law of thermodynamics that the absolute zero of temperature can never be reached.

The surface temperatures of stars range from a few thousand to more than 40,000 K (see box for definition of this temperature scale). Every object having temperature above absolute zero (which is the lowest possible temperature anywhere in the Universe) has *thermal energy*. This is mostly in the form of kinetic energy of particles constituting the object. In gases and liquids, the motion of the particles is random as the particles are free to move around. In solids, the thermal energy is in the form of vibrations of particles around their equilibrium positions. In addition, the thermal

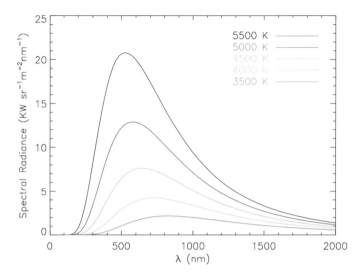

Fig. 2.5 The spectrum of radiation emitted by a black body for five different temperatures. The horizontal axis indicates the wavelength of the radiation while the vertical axis indicates the energy in the radiation per unit wavelength. Image credit: Kaushal Sharma

energy can be in the form of excitation energy, i.e. some of the atoms could be in an excited state and have thermal energy stored as excitation energy. The higher the temperature of an object, the higher is its thermal energy. Every object having thermal energy emits electromagnetic radiation. This is called the thermal radiation as it is on account of and at the expense of the thermal energy possessed by the body.

The magnitude of thermal radiation emitted by a body at different wavelengths in general, depends on several properties of the body like the temperature, size, shape, composition, etc. However, there exists a class of objects called 'black bodies' which have a characteristic spectrum; the shape of the spectra emitted by these bodies (i.e. the relative magnitudes of the radiation emitted as a function of frequency or wavelength) depends only on the temperature of the body. The total amount of energy emitted by a black body also depends on its surface area. The radiation emitted by stars can be approximated to be that of a black body.

The amount of energy emitted by a black body at different wavelengths, for five different temperatures of the black body, is shown in Fig. 2.5. Each curve is known as a black body spectrum. The total amount of energy emitted by the body is equal to the area under the curve. It can be seen that higher the temperature of the body, the higher is the total amount of radiation emitted, and lower is the wavelength at which maximum energy is emitted. We actually see this in our daily life. Remember that an iron rod when heated changes its colour. As its temperature rises its colour changes in the following order: black, red, orange, yellow, white and bluish white. This is because, the rod actually behaves like a black body. As the temperature of the rod increases, it emits more and more energy in radiation of smaller and smaller wavelengths. This is responsible for the change in colour perceived by us.

What are the properties of a body which decide whether or not it is a black body? If a body is in thermodynamic equilibrium at a given temperature, i.e. the amount of energy being emitted by a body is equal to the heat gained by the body, then in the ideal case, the spectrum of the radiation emitted by it is that of a black body. But in reality, the measured spectrum is hardly ever a perfect black body spectrum. The distortions occur because not all parts of the body may be at the same temperature, so that the radiation is a mixture of black body spectra at different temperatures. The spectrum can also be distorted because various emission and absorption processes add energy at some wavelengths and remove energy from other wavelengths. We will see examples of such distortions when we consider the energy spectrum of stars.

2.4 Special Theory of Relativity

The *special theory of relativity* was published by Albert Einstein in 1905. Through special relativity, Einstein completely revolutionised the ideas of space and time. He proposed that the three-dimensional space that we live in, and time, together constitute a four-dimensional entity called space-time. The theory was based on two postulates, as described below, and all physical laws had to be consistent with the theory. While the ideas were abstract, they had very important consequences, many of which were verified in the following decades. Special relativity and quantum mechanics, to the development of which too Einstein made seminal contributions, form the basis of modern physics.

2.4.1 The Velocity of Light

The speed of light in vacuum is 300,000 km/s. As described in Sect. 2.3.1, according to Maxwell's theory of electromagnetism, light consists of electromagnetic waves, which propagate through an all pervading ether which is taken to be at absolute rest. The speed of electromagnetic waves was believed to be 300,000 km/s only when measured by an apparatus which is at rest with respect to the ether. The value would be different if the observer, and therefore the apparatus that he is using to measure the velocity, is moving with respect to the ether. The reason for this can be easily understood by considering the following example.

2.4.2   The Addition of Velocities

Imagine two boys A and B playing with a ball. In the first case, both are stationary with respect to each other and B throws the ball to A with a speed of 20 km/h. Obviously, A will find the ball coming to him at 20 km/h. This is shown in Fig. 2.6a. Now imagine A standing on a cart which is being pushed towards B with a speed of 5 km/h while catching the ball thrown by B towards him as shown in Fig. 2.6b. The speed at which the ball will be received by A will be $20 + 5 = 25$ km/h. If, on the other hand, the cart is being pulled away from B at the same speed (i.e. 5 km/h) and B throws the ball towards him with the same speed of 20 km/h then A will receive it with the speed of $20 - 5 = 15$ km/h. Thus, the speed of the receiver (here A) gets either added (or subtracted) to the speed of the ball if he is moving towards (or away from) the ball while receiving it. Yet another situation will be if the cart is being pulled in a direction perpendicular to the direction of the ball's travel and the velocity of the ball is perpendicular to the velocity of A, at the instant that he catches the ball. In this case, the velocity at which the cart is pulled will have no effect on the velocity of the ball along its direction of motion relative to the ground, which A will perceive as 20 km/h, as shown in Fig. 2.6c. The total velocity of the ball as perceived by A will also have a component along the direction of motion of A (of course be different, and will be about 20.6 km/h). Thus, we can see that the velocities of the source and the apparatus (in this case the two boys B and A, respectively) along the line joining them get added or subtracted, to or from the velocity of the ball, depending on the direction of motion.

2.4.3   Measuring the Effect of Motion of the Detector with Respect to the Ether on the Velocity of Light

If a moving apparatus is used to measure the speed of light, it should measure a different speed as compared to a stationary apparatus, because it would be moving with respect to the ether. In the nineteenth century, the velocities achievable for any apparatus on the Earth were very small compared to the velocity of light, and it was not possible to perform experiments to measure the change in velocity of light due to this motion as it would be too small to be measured. There was, however, a way out: a large velocity was naturally present in the form of the velocity of motion of the Earth as it moves around the Sun, which is about 30 km/s. This is high enough for a test to be made. We can compare this situation with the simple example given in Sect. 2.4.2. The cart there was moving relative to the ground, which was taken to be fixed. The velocities for the balls as measured by A were different in two cases, i.e. (i) when the cart he stood on was being pulled in a direction perpendicular to the direction of motion of the ball (Fig. 2.6c), 20 km/h and (ii) when the cart was moving towards or away from B (Fig. 2.6b), 25 or 15 km/h, respectively. In the same manner, the velocity of light as measured by an apparatus stationary with respect to

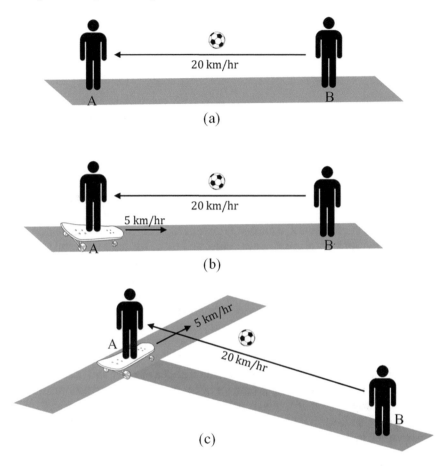

Fig. 2.6 The velocities of a ball thrown by B and received by A in three cases, **a** where both are stationary with respect to each other, **b** where A is moving towards B with a speed of 5 km/hr and **c** where A is moving perpendicular to the direction of travel of the ball with a speed of 5 km/hr. Image credit: Kaushal Sharma

the Earth but moving with it through the ether should depend on which direction the apparatus is moving with respect to the ether.

Two American physicists, Albert A. Michelson and Edward W. Morley, performed such an experiment in 1887. With their apparatus, they measured the velocity of light in two directions: one in the direction of Earth's motion through the ether and the other perpendicular to this motion. The two values were expected to be different similar to the velocities of the balls received by A in the two cases mentioned above. To their dismay, they found the velocity of light to be exactly the same in both directions! The implication was that motion of the detector relative to the ether was not affecting the measured speed of electromagnetic waves. This went against the expectation from Maxwell's otherwise very satisfactory theory. Physicists were at a loss to understand the reason for this failure and the ether theory remained an enigma for some years. While some suggestions for a way out of the difficulty were proposed, none proved to be wholly satisfactory.

2.4.4 Postulates of Einstein's Special Theory of Relativity

The problem of the ether was finally solved by the development of special relativity by Albert Einstein, which he announced in 1905. Einstein was motivated by considerations which went beyond the *Michelson-Morley experiment*. He noticed that experimentally, only relative motion was important. As an example of this, we can consider Faraday's law. If we hold a loop of wire at a fixed place and move a magnet back and forth through it, a current is generated in the loop. The same result is obtained if we hold the magnet in one place, and move the loop back and forth. It does not matter whether the magnet moves or the loop moves or both move. What is important is that the two move relative to each other. On the other hand, in Maxwell's theory, the situation in which the magnet moved had a different interpretation from the situation in which the loop moved, even though both situations produced the same result, i.e. generation of current. Einstein realised that the problems arose due to faulty understanding of the concepts of space and time, and that a new theoretical framework was required for physics.

Einstein took the radical step of discarding the idea of ether and made the bold postulate that the velocity of light (and therefore of all electromagnetic waves) in vacuum is always the same and does not depend on the velocity of the observer or the source. What is the implication? Imagine a person standing on the ground who shines a torch in the direction of motion of the Earth. According to the ether theory, the velocity of the light beam relative to the ether (which is at rest) should be 300,030 km/s, which is the sum of the speed of light and the motion of the Earth through the ether (which is exactly what Michelson and Morley had aimed to measure). But Einstein asserted that the velocity of the beam must be just 300,000 km/s, which seems absurd, but was in accord with the result of the Michelson-Morley experiment even though Einstein did not mention that fact in his paper.

To complete the basis of special relativity, Einstein needed a second postulate. In physics, there is a special type of observer known as an inertial observer. For such an observer, the laws of physics take their simplest form, and an object, which has no force acting on it, remains at rest or moves in a straight line with constant speed. Einstein's second postulate is that all observers who are at rest or are moving with constant velocity with respect to an inertial observer are themselves inertial observers. Hence, the laws of physics hold good and take the same mathematical form for all such observers.

2.4.5 Some Interesting Consequences of Einstein's Theory

Einstein's two postulates lead to the unification of space and time into the single entity of space-time, as mentioned above. As a consequence, special relativity predicts many interesting consequences of high-speed motion. Some of these are (i) a clock moving with respect to an observer appears to go slower than a clock which is at rest with respect to her; (ii) a rod moving with respect to an observer appears to shrink in length when compared to a similar rod at rest with respect to her; (iii) two events which appear to occur at the same time, i.e. simultaneously, to one observer may not appear to be simultaneous to another observer moving with respect to her and (iv) the mass of a particle in motion is greater than its mass when it is at rest. The most famous consequence, of course, is the equivalence of mass and energy, which is expressed through the equation $E = Mc^2$, which is familiar to people of every background. In special relativity, the simple law of addition of velocities as discussed in Sect. 2.4.2 does not hold, and a more complex relation is needed. The new law leads to the dramatic conclusion that nothing, no material particle, no wave, no energy in any form and so no information in the Universe, can travel faster than the velocity of light in vacuum which is 300,000 km/s.

The equations of mechanics describing motion had to change due to special relativity, Maxwell's equations were found to be fully compatible with it. But the interpretation of observed phenomena is changed; there is no ether and the speed of all electromagnetic waves in vacuum is the same for all inertial observers.

The creation of the special theory of relativity by Einstein was a remarkable flight of mind and is one of the reasons contributing to Einstein's being said to have possessed the greatest brain that ever existed. However, it was difficult to verify Einstein's theory in the laboratory as very large velocities are required to test the new law of addition of velocities, and other predictions, because of the large value of the velocity of light. However, it was eventually done in many cases, mostly using natural phenomena. The discovery of high-energy cosmic rays allowed the testing of length shortening and the slowing of clocks. Einstein's special theory of relativity now stands fully verified.

Twin Paradox

One result of Einstein's special relativity is the twin paradox. According to his theory, a moving clock appears to go slow compared to a stationary clock. Let us assume that two clocks are stationary and are synchronised. If one of them which we will label as B starts moving with respect to the other, let us call it A, then it will run slower than A. If B returns to A after a while, then it will show time which is earlier than the time shown by A. Einstein presented this in another way. Suppose one of a pair of twins, say A, remains stationary while the other, say B, moves at great speed for a while and then returns to A, then his age will be less than that of A.

The reason this is called a paradox is that velocity is relative and if B was moving with respect to A, then from B's point of view, A was moving with respect to him and should be younger than him when they reunite. So who will be younger? The answer is that B was not only moving, but was accelerated twice, once on his way away from A and second when he turned back to return to A. A was, however, stationary and did not undergo any acceleration. So we can differentiate between the motions of A and B and hence the age of B will be less and there is no real paradox.

It is interesting to know that NASA has recently conducted a similar experiment. It sent one of a pair of twins to the space station where he spent an year, the other member of the pair remaining on the Earth. After reuniting, their health parameters were compared and some conclusions have been drawn from the study.

2.5   Learning About the Universe Using Electromagnetic Waves

The electromagnetic radiation received from an astronomical source is loaded with information about the source. There are several properties of radiation that help us decipher the nature of the source. A few of these pertaining to stars are described below. Along with visible radiation, some of the cosmic objects also emit copious amounts of electromagnetic radiation at various wavelengths, including in the radio, infrared, ultraviolet, X-ray and Gamma ray regions. Such radiations provide us information about high-energy phenomena occurring in the sources. We will not consider them further in this section, and describe them as needed in other parts of the book.

The matter in stars is in a gaseous form because of the high temperatures prevailing in them, ranging from several thousand Kelvin at the surface to tens of millions of Kelvin in the central regions. The gas consists mostly of hydrogen and helium with small amounts of heavier elements. The energy emitted by stars is generated at their centres by nuclear reactions taking place there. We will learn more about that in Chap. 5. The energy generated at the centre is in the form of very high energy radiation, i.e. in the form of Gamma rays. These cannot come out of the star directly as they get absorbed by the stellar material on travelling a small distance away from the centre. They deposit their energy there and so heat up the gas. The temperature of the gas there is somewhat less than the temperature at the centre. It emits thermal radiation of average frequency smaller than that emitted by the gas at the centre. This gets absorbed on travelling a further short distance towards the surface and deposits its energy, thereby heating the gas there. The radiation in this manner is repeatedly absorbed and reemitted on its way out to the surface, and a temperature gradient is set up in the star. The temperature at the centre, where nuclear reactions take place,

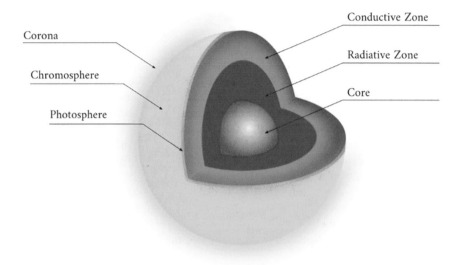

Fig. 2.7 Some regions of a star. The central core where energy generation takes place is shown in pink. Also shown are the photosphere, chromosphere and corona. The convective zone and radiative zone are regions in which the transfer of heat occurs through convection and radiation, respectively. Image credit: Kaushal Sharma

is the highest and at the surface the temperature is the lowest. For the Sun, the central temperature is about 15 million K while that at the surface is about 5800 K.

What exactly do we mean by the surface of a star? This is the outward layer of the star, the radiation emitted by which can leave the star directly, as there does not exist too much gas beyond it to absorb the radiation. This surface layer is called the photosphere, as this is the luminous sphere that we see when we look at a star. Some cooler material does exist outside the photosphere and is part of the atmosphere of the star. The layer immediately outside the photosphere is called the *chromosphere*, and the layer beyond the chromosphere is called the *corona*. These layers are schematically shown in Fig. 2.7. The temperature keeps decreasing from centre to some point in the chromosphere where it reaches close to 4300 K for the Sun, and then increases again, up to three million K in the corona of the Sun.

In the next subsection, we look at some of the properties of the star that astronomers measure from the radiation received from it.

2.5.1 Luminosity

Th *luminosity* of a star, which is the amount of radiation emitted by it per second, is an important property of a star (or any other astronomical source). How bright the star appears to us, i.e. its *apparent* brightness depends on its luminosity as well as its distance from us. The star may be emitting a great deal of radiation, i.e. it may be having very large luminosity but if it is very far from us, it appear to be faint to us. This is similar to the case of street lights. A distant street light appears fainter than a nearer one even though they both have identical wattages. On the other hand, an astronomical source may not be intrinsically very bright but may be close to us and so may appear very bright. This is the case for the Sun, which is an average star. There are many stars which emit much more radiation than the Sun, but they appear faint because they are much further than the Sun. So from the apparent brightness of the source, we have to calculate the actual amount of energy radiated by the source by knowing its distance. For example, in the case street lights, if we measure the apparent brightness of a distant street light and we know its luminosity (because we know its wattage), we can calculate its distance from us; vice versa, knowing the distance and measuring the apparent brightness of a distant street light, we can determine its luminosity. Thus, measuring any two out of the (i) apparent brightness, (ii) luminosity and (iii) distance, we can determine the third property of the star.

2.5.2   Distance

The distance to an astronomical source can be measured by a direct method if the source is sufficiently near to us and by indirect methods for sources which are far away. The direct method can be used if the sources are closer than about 15,000 light years, where a light year is a unit that astronomers use to measure large distances. One light year is the distance travelled by light in 1 year and is equal to 9.46 trillion km. This method for measuring distance uses the properties of an isosceles triangle. The direction of the source is measured from a point on the Earth at a time gap of 6 months. At those two times, the Earth is at diametrically opposite points on its orbit around the Sun. The distance between these two points is the base of the triangle. This is shown in Fig. 2.8. The astronomical source is at the vertex of the triangle and the directions of the star from these two points give the angle that each side of the triangle makes with the line joining the Sun and the star and is known as the *parallax*. Knowing the angle, and the radius of the orbit, which is about 150 million km, the height of the triangle, which is the distance to the source, can be determined. This is known as the parallax method.

As the distance to a star increases, the parallax becomes too small to be measured accurately, in which case other methods, which depend, for example, on some physical properties of the stars, are used. In such indirect methods, use is made of properties of some specific class of sources which have a direct relationship between their luminosity and some particular property. For example, there are variable stars known as *Cepheid variables* whose luminosity periodically varies with time with a certain recognisable pattern. The period of variation is directly proportional to the luminosity of the star. The period of variation in luminosity can be measured even if a star is very far away. From the period, one can get the luminosity of the star using the relation between the two. This along with the apparent brightness, which

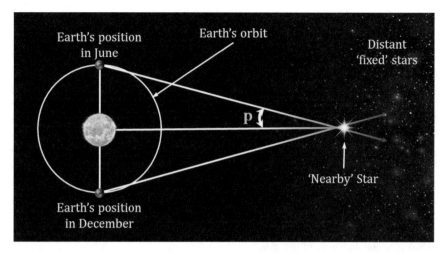

Fig. 2.8 The parallax method of measuring the distance to nearby astronomical sources. The position of a nearby star, relative to more distant stars, is measured from diametrically opposite points along the Earth's orbit. The measured parallax angle p provides a measure of the distance as described in the text. Image credit: Kaushal Sharma

again can be measured, can be used to determine the distance to the star as seen in Sect. 2.5.1. Other sources for which a distance determination method is available are supernova (Type Ia) explosions. We will learn more about these sources in Chap. 5. Here it is sufficient to note that the luminosity of these exploding stars decides the rate at which its luminosity or apparent brightness changes with time over a period of days. Observing the rate of change of apparent brightness can give us the luminosity of the source and thereby its distance.

2.5.3   Temperature

One can determine the temperature of a star from its spectrum. We have seen in Sect. 2.3.3 that a star emits like a black body, and therefore its temperature determines the shape of its spectrum. Thus, one can determine the temperature of a star by obtaining its spectrum and matching it with that of a black body having suitable temperature. The method can be made more accurate by taking into account the detailed shape of the spectrum including the absorption lines and other features.

2.5.4 Size

One can determine the size of a star directly if it is sufficiently near to us so that one sees it as a disc and can measure (i) its *angular diameter*, which is the angle subtended by two diametrically opposite points of the star at our position and (ii) its distance. This can be done by using properties of a triangle whose two sides are equal to the distance of the star and the third side is the diameter of the star. This is exactly the method used to estimate the height of a mountain from its apparent height, which is the angle subtended by the mountain at our eye, and its distance from us. If a star is too far and appears like a point source this method cannot be used for measuring its size. A simple method for estimating the size of a star is using its luminosity and temperature. The luminosity is known if its distance is known or it can be inferred using indirect methods. The temperature can be estimated as described above. It is then a simple matter to use laws related to black body emission to estimate the diameter of the star.

2.5.5 Chemical Composition

The dark lines, called absorption lines, in the spectrum of astronomical sources seen in Fig. 2.9 are basically finger prints of the material present above the surface, i.e. in the atmosphere, of the star. These lines are produced because ions, atoms and molecules of various elements in the cooler material in the atmosphere of the star absorb radiation of specific wavelengths coming from the star's inner region. An electron in an atom of a particular element makes a transition from lower energy state to a higher energy state by absorbing radiation (photons) having energy equal to the difference in the energies of two energy levels. The wavelength of the radiation absorbed by an atom is thus determined by the structure of the atom. If there are sufficient number of such transitions, a significant amount of radiation of that wavelength is absorbed and hence removed from the radiation coming out of the star and we receive less radiation at those wavelengths as compared to the neighbouring wavelengths. This shows up as dark lines at these wavelengths in the spectrum. The larger the number of atoms or ions making such transitions, the larger will be the amount of radiation absorbed and darker (stronger) will be the absorption line. For example, in Fig. 2.9, some lines due to iron (marked as Fe, which is the chemical name for iron) and a line due to magnesium (marked as Mg, which is the chemical name for magnesium) are shown. The rightmost line marked as Fe is darker than the other lines, indicating greater absorption at that wavelength. By detailed measurements of the wavelengths and strengths of the absorption lines and further analysis of the spectrum, scientists can determine which chemical elements are present in the atmospheres of stars, in what form and in what amounts.

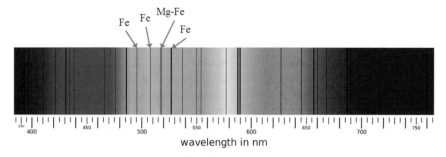

Fig. 2.9 The spectrum of the Sun produced using a spectrograph. The dark lines seen in the spectrum, which are called *Fraunhofer lines* after their discoverer, are produced by chemical elements present in the region close to the surface. Some lines produced by iron (chemical name Fe) and magnesium (chemical name Mg) are shown. Image Courtesy: Kaushal Sharma

Helium

In 1868, certain dark lines were discovered in the spectrum of the Sun. These did not seem to match the fingerprints, i.e. wavelengths, of lines produced by any of the chemical elements known to be present on the Earth. Scientists, therefore, concluded that the lines were produced by an element not known on the Earth. This newly discovered element was named helium (Helios meaning the Sun). It was later in 1895 that helium was discovered on the Earth.

2.5.6 Radial Velocity, Mass and Magnetic Field

Another important information encrypted in the spectrum of a star is the knowledge about its velocity towards or away from us which is known as its *radial velocity*. This is because of what is called the Doppler effect which holds good for all kinds of waves and is routinely experienced by us for the case of sound waves. The sound of the whistle of a train or that of the siren of an ambulance passing us is different when the vehicle is approaching us as compared to when it is receding from us. This happens because the wavelength of a sound wave emitted by a source appears to be shorter than the actual wavelength of the emitted wave, when the source is approaching us. Similarly, the wavelength appears to be longer when the source is receding from us. The wavelength of the sound wave emitted by the source (train or siren) does not change, it is only its value as perceived/measured by us that changes. The pitch, i.e. the frequency of the sound received by us is thus different in the two cases; higher when the source is approaching and lower when it is receding. This phenomenon happens for all kinds of waves, including the electromagnetic waves. This increase and decrease of wavelength of light, and electromagnetic waves, in general, is called *redshift*, i.e. shift towards red and blue shift, i.e. shift towards blue, respectively. Thus, if we know the wavelength of the radiation that is emitted by the source and measure the wavelength of the radiation reaching us, we can determine whether the source is approaching or receding with respect to us and with what velocity.

More than half of the stars in our *galaxy* are present in *binary systems*. A binary system is a system of two stars which are going around, i.e. revolving around each other. By studying the luminosity and spectrum of such stars one can determine their orbits and velocities and then obtain their masses using *Newton's law of gravity*. There is much more information hidden in the spectrum than we can describe here and by using it astronomers can gain further information about the sources, e.g. the strength of the magnetic field, ages of the stars and so on.

Thus, it can be seen that the radiation coming from the astronomical sources is a treasure trove of information about the sources and by careful and detailed observations and analysis astronomers try to extract as much of it as possible. Electromagnetic radiation has been the main source of our knowledge of the Universe.

Chapter 3
Gravity: The Force that Governs the Universe

Abstract In this chapter, we describe the phenomenon of gravitation. It is the weakest of the four known fundamental forces, and yet it governs the nature of astrophysical objects like stars and galaxies, and the structure and evolution of the whole Universe. Gravity was first recognised as a universal force between any two particles of matter by Isaac Newton in the seventeenth century. Using his famous law of gravitation and the calculus that he independently developed, Newton was able to describe the motion of objects near the surface of the Earth, the orbit of the Moon around the Earth and the orbits of planets around the Sun. In spite of all its successes, Newton's theory is not consistent with the special theory of relativity. That led Albert Einstein to develop an astonishing new theory, known as the general theory of relativity, which describes gravity as the expression of curvature in space-time, produced by matter and energy. Einstein's theory is compatible with special relativity and it agrees with Newton's theory when gravity is weak as in the Solar system. It is able to account for a small discrepancy in the observed orbit of the planet Mercury compared to the orbit predicted by Newton's theory, predict the correct value for the bending of light due to gravity and predict the existence of gravitational waves. The gravitational bending of light is at the basis of the gravitational lensing, which has proved to be an important tool for astronomers.

3.1 Introduction

Gravity is the universal attractive force which acts between all bodies in the Universe. When we throw a ball up into the air, it rises to a certain height, and then stops and returns to the Earth. It does so because the Earth exerts a gravitational force on the ball, reducing the ball's speed until it stops and then increasing the ball's speed in the downward direction, making it fall back to the Earth. The same gravitational force acts between the Earth and the Moon, between the Sun and the Earth, and between our Milky Way galaxy and all other galaxies in the Universe.

That such a universal force exists was first realised by the great English physicist Sir Isaac Newton. There is a famous story about Newton seeing an apple fall from a tree in his garden in Woolsthorpe Manor, his birthplace, in Lincolnshire in England.

The sight is believed to have led him to think about a universal force that may be existing between all bodies. In this chapter, we will describe Newton's theory of gravity and its importance for astronomical phenomena. Then we will describe how Albert Einstein viewed gravity in his *general theory of relativity*, and in a later chapter describe how his theory predicts the existence of gravitational waves.

Newton
Isaac Newton (1642–1726) was born in England. His father died 3 months before his birth. His mother remarried when he was 3 years old and entrusted him to the care of his grandmother. Because of this, Newton always held a grudge against his mother and step father. He took admission to Trinity College in Cambridge and paid his fees by working as a valet until 1664 when he was awarded a scholarship. The college closed down for 2 years in 1665 because of a plague epidemic and Newton was forced to study at home. He did his best work in those years. This included his work on light and on gravitation. During this time he also discovered independently a branch of mathematics which is known as calculus. He is recognised as one of the greatest physicists and mathematicians of all times because of his fundamental and important contributions to these fields. But in his own words he felt that 'I do not know what I may appear to the world, but to myself I seem to have been only like a boy playing on the sea shore, and diverting myself now and then finding a smoother pebble or a prettier shell than ordinary, while the great ocean of truth lay all undiscovered before me'.

3.2   The Four Fundamental Forces

How one body influences the motion of another body depends on the forces they exert on each other. For example, an apple on a tree falls to the Earth because of the gravitational force of attraction between them. We have seen in the last chapter how electrically charged bodies can exert a force on each other. Are these the only two forces in nature? We know that all the bodies are ultimately made up of atoms, which themselves contain elementary particles called protons, electrons and neutrons. So we should be able to understand the action of such bodies on one another, if we can understand all the forces that act between the particles that make up the bodies.

From studies which have spanned about 300 years from Newton's discovery of gravitation, to Maxwell's work on electromagnetism, to the development of atomic and elementary particle physics in the twentieth century, physicists have determined that there are only four basic or *fundamental forces* in nature. These forces, in order of increasing strength, are (1) gravitational force, (2) weak force, (3) electromagnetic force and (4) strong force. We have already learnt in the last chapter about the electromagnetic force, and phenomena related to it. In this chapter, we will learn about the gravitational force.

Discovered in the twentieth century, the weak and the strong forces act only over very short distances, about 10^{-15} m (one thousandth of one millionth, millionth of a metre). This distance is smaller than the typical size of the nucleus of an atom. These forces are extremely important in phenomena related to elementary particles, atomic nuclei and nuclear reactions. They play a central role in nuclear energy generation in nuclear reactors, in atomic weapons and also inside stars. We will not be dealing with these two forces in this book. Unlike the weak and strong forces which operate only over atomic dimensions, both the gravitational and electromagnetic forces can be felt at very large distances, though the magnitude of these two forces reduces in the same manner with increasing distance.

3.2.1 The Importance of the Gravitational Force

In spite of being the weakest by far of the four forces of nature, the gravitational force governs the nature of bodies of very great mass, like planets, stars and galaxies, and the structure and evolution (change with time) of the Universe taken as a whole. Why is gravity so important? Firstly, the weak and the strong forces are felt only at very small distances, while most of the heavenly objects in the Universe like the planets, stars, galaxies, etc are situated at very large distances from one another (millions of km to hundreds of millions of light years). Secondly, even though the electrical force can be felt at large distances, it is exerted only between charged bodies. All heavenly objects have the same number of positively and negatively charged particles, i.e. protons and electrons, respectively, and hence are electrically neutral. Thus, there is no net electric force between them. A magnetic field is present in most places in the Universe but its magnitude is very small and it too does not have a net force on electrically neutral bodies, and hence the magnetic force is also not important.

Of the four fundamental forces, we are therefore left only with gravity to act between the very massive bodies over very great distances. The force of gravity is always attractive, so there is a net resultant attractive force between any two bodies. Also, in spite of the large distances between heavenly bodies, the gravitational force is substantial due to their huge masses. We therefore need to understand gravity in order to understand how the Universe functions.

The role of gravity is seen in all astronomical phenomena. In the Solar system, it is responsible for the occurrence of tides, for the path followed by the comets, and for the revolution of planets in their fixed orbits for billions of years around the Sun. Gravity is responsible for the formation of stars and planets from huge and rarefied interstellar clouds and for the spherical shape of the stars and planets. It is the force that determines the formation and stability of galaxies, and the expansion and the ultimate fate of the Universe. Let us learn more about this important force.

3.3   Newton's Theory of Gravity

The gravitational force was postulated and understood to a large extent by Newton in the second half of the seventeenth century. He was of course aided in his efforts by the earlier work of Nicolaus Copernicus, Tycho Brahe and Johannes Kepler on planetary motions. We have mentioned above the sight of an apple falling from a tree prompting Newton to postulate the universal force of gravity. The story is indeed true and Newton himself told of it to a friend who later wrote it up in a biographical work on Newton. Seeing the apple fall, Newton wondered why apples always fall vertically down and do not move horizontally, or at an angle with respect to the vertical. He then realised that it must be the matter in the Earth that is attracting the apple. He worked out that the attractive force due to the Earth must be directed towards its centre. Thus, the apple falls towards the centre of the Earth, or vertically downwards as we see it. He later wondered if the force due to the Earth extends much farther than the height of the apple tree, perhaps all the way to the Moon. These thoughts prompted him to postulate the universal gravitational force.

That the Earth attracts bodies towards itself had been known before Newton and even Brahmagupta, the ancient Indian astronomer and mathematician, is credited to have said so in the ninth century. The genius of Newton was in postulating that the force between an apple and the Earth is exactly the same in nature as the force which is responsible for the motion of the Moon around the Earth, and that of the planets around the Sun. His law is called *Newton's law* of universal gravitation. According to this law, every particle in the Universe attracts every other particle. The force of attraction, called the gravitational force, depends on the masses of the two particles and the distance between them. The dependence is such that the force is greater for higher masses and smaller for greater distances. If the distance doubles, the force decreases to a quarter of its value, if the distance triples, then the force decreases to a ninth of its value and so forth. The force on the apple on the tree in Fig. 3.1 is roughly 3650 times larger than that on an identical apple kept on the surface of the Moon. Mathematically, one would say that the force is proportional to the product of the two masses, and inversely proportional to the square of the distance. The statement of the law also includes Newton's gravitational constant. The dependence on distance is similar to that in Coulomb's law for the force between two charges.

Newton was able to prove that the gravitational force due to a spherical object on a particle should be directed towards the centre of the spherical body. The Earth is very nearly spherical in shape, so the force due to the Earth on an apple should be directed towards the centre of the Earth. That is also true with the gravitational force of the Earth on the Moon, as shown in Fig. 3.1. The two forces in effect are perpendicular to the surface of the Earth from the positions of the apple and the Moon, respectively.

As every body gravitationally attracts every other body, the apple also attracts the Earth. But the Earth being very massive moves very little and in effect it is the lighter apple that falls towards the Earth. The Earth and Moon attract each other, but the

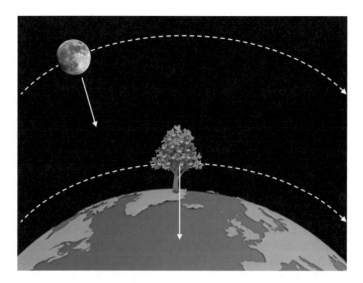

Fig. 3.1 The directions of the gravitational force due to the Earth on an apple on a tree and on the Moon. The two forces are directed towards the centre of the Earth. The distances shown are not to scale. Image credit: Kaushal Sharma

Moon does not fall on the Earth like the apple does, because it is moving in an orbit around the Earth. If it stops moving it will immediately fall on the Earth.

One important property of the gravitational force is that it is always attractive and never repulsive. Thus, all masses always attract each other because of this force. Another property is that the nature of the masses is totally unimportant. Thus, the force of attraction between an iron ball and a gold ball will be same as that between blobs of water and of jam if they have the same masses and the same distance between them.

3.3.1 Successes of Newton's Theory

Newton's theory was highly successful in mathematically explaining the motion of objects thrown on the Earth, motion of the Moon around the Earth, as well as the motion of planets around the Sun. The observations of the positions of the planets had been collected over long periods of time. These included very detailed and accurate observations, particularly on the orbit of the planet Mars, made by Tycho Brahe. Johannes Kepler, who began as a research assistant of Brahe, thoroughly studied observational data on Mars during the first two decades of the seventeenth century. He eventually derived three laws on planetary motion based purely on the observations. These essentially tell us how the planets move around the Sun and how their motion depends on their distance from the Sun. But he did not know why planets follow these laws.

Newton mathematically proved these laws using his theory of gravitation and his laws of motion. For calculations like these, Newton needed a mathematical discipline known as Calculus, which he developed independently of other efforts. His studies of calculus alone make Newton one of the greatest of all mathematicians.

A spectacular success of Newton's theory came around the middle of the nineteenth century. The seventh planet of the Solar system, Uranus, had been discovered in 1781 by John Herschel by a chance observation. By the early 1840s it was close to completing one revolution around the Sun after its discovery. Scientists found that its orbit did not seem to be as predicted by Newton's theory of gravitation, and there were some measurable deviations from the predictions. Could Newton's theory be wrong? That seemed unlikely, considering all the success it had met untill then. It was then realised that the irregularities in the orbit of Uranus could arise if there was an eighth planet present in the Solar system, farther from the Sun than Uranus, which was gravitationally influencing the motion of Uranus. In 1845, using Newton's theory, scientists calculated the position of the hypothetical planet which would have the required effect on the orbit of Uranus. A search was made for the hypothetical planet using telescopes and a planet was actually found to be present at the predicted position in 1846, which was named as Neptune. This was indeed a great success of Newton's theory.

Newton used his theory to describe the slight flattening of the Earth at the two poles as the combined effect of its own gravity and its rotation. He was also able to describe the phenomenon of ocean tides as a consequence of the gravitational attraction of the Sun and the Moon on the water in the oceans. Edmund Halley used Newton's theory to predict that certain comet appearances which had been observed over the centuries were actually just one comet which returned to the vicinity of the Earth about once every 76 years. He preformed calculations of its orbit to predict that the comet would appear again in 1758. The comet, subsequently named after Halley, returned as predicted, though Halley did not live to see the reappearance. In modern times, Newton' theory of gravity is used to determine the orbits of artificial satellites, the structure of stars like the Sun, the shapes of galaxies and in astrophysical calculations, in general, when the force of gravity is not very large, as explained below.

3.3.2 Limitations of Newton's Theory

Even though Newton could explain the motion of planets around the Sun, he was extremely uncomfortable with deeper implications of his theory of gravitation. We see in day-to-day life that we can exert a force on an object, i.e. push or pull an object only when we touch it either directly or indirectly using a rope, stick, etc. But according to Newton's theory, two bodies exert gravitational force on each other even when they are not touching and are very far, and even when they are present in vacuum, without any medium between them. Also, a change in mass or position of one object is instantaneously felt by another object, a long distance away, as a change in the gravitational force that it experiences towards the first object. This is called instantaneous action at a distance. Newton wrote in his letters to a friend that it is highly absurd that the gravitational force will be conveyed from one body to another at large distances from it without any agent or medium. He could not find any answer to this dilemma.

After Einstein developed his special theory of relativity (Sect. 2.4), the problem with instantaneous action at a distance in Newton's theory became particularly acute. According to the special theory, no body or information can travel faster than the velocity of light in vacuum. However, in Newton's theory, the gravitational force of an object is instantaneously felt by another object, however far it may be. Thus, the gravitational influence travels with velocity faster than that of light, in fact with infinitely large velocity, which is a contradiction with the special theory of relativity.

Another problem with Newton's theory was discovered by a French Mathematician, Le Verrier, in 1859 as a minor discrepancy between its prediction and actual observation. According to Newton's theory, the planets move in elliptical orbits around the Sun. If there was only one planet moving around the Sun, the orbit would be an ellipse which would remain fixed in space. However there are other planets in the Solar system which exert gravitational forces on one another. These additional forces cause the ellipse to precess, that is the whole ellipse rotates around the Sun. Thus, the point along the orbit of a planet when it is closest to the Sun, which is known as the *perihelion* does not remain fixed in space but shifts by a small amount during every complete revolution around the Sun. This is shown in Fig. 3.2. Using Newton's theory, the change in this position can be calculated. The planet Mercury is closest to the Sun, and for it the *precession* is the greatest. The calculated value of the angle between two such positions of nearest approach which are 100 years apart, as seen from the Sun (see the figure), for planet Mercury is 532 arcseconds (1 arcsecond = 1/3600 of a degree). But observations found this angle to be 575 *arcseconds* per century. So there is a discrepancy of 43 arcseconds per century between the observations and Newton's theory. This was a tiny amount, but nevertheless it pointed to an observational flaw in the theory.

With the discovery of massive very compact astrophysical objects like neutron stars and *black holes* in the second half of the twentieth century, it was found that Newton's theory was inadequate to describe such objects which had very strong gravitational fields. Newton' theory also was not able to fully describe the expansion of the Universe discovered independently by Georges Lemaître and Edwin Hubble in the 1930s (see the Boxes Galaxies and *Cosmological Redshift* in Sect. 5.4.3.1). Einstein's theory of gravity is needed to explain such objects and phenomena.

The successes of Newton's theory were too numerous to discard it because of the problem with precession. It was left to the genius of Einstein to formulate a new theory of gravity which could account for this discrepancy. This new theory was consistent with the special theory of relativity and in fact was broader than it. The theory resolved the dilemma faced by Newton regarding action at a distance and explained the observed discrepancy in the precession of the perihelion of Mercury, while keeping other predictions of Newton's theory intact. Like all good theories, it made important new predictions, many of which have been proven to be correct.

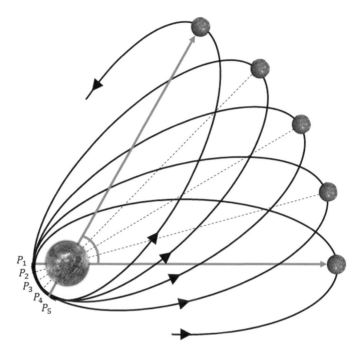

Fig. 3.2 The orbit of Mercury and the change with time in the perihelion, which is the position of Mercury when it is closest to the Sun. P1, P2,..P5 are successive points of closest distance. The change in the direction of perihelion over five orbits is shown in red. Over a century, which corresponds to about 415 orbits, the change in direction is 575 arcseconds. Image credit: Kaushal Sharma

3.4   General Theory of Relativity

The new theory of gravity developed by Albert Einstein is called the General Theory of Relativity. He worked on it for 10 years after he discovered the Special Theory of Relativity, and finally completed it in 1915. The theory had very novel ideas, required complex mathematics and was extremely difficult to comprehend. This can be seen from the quote of the famous astronomer Sir Arthur Eddington. When asked in 1919 (4 years after Einstein gave the theory) whether it was true that only three persons in the world understood the theory, he asked 'who is the third?', he himself being the second person, Einstein of course being the first.

We described Einstein's special relativity briefly in Chap. 2. In this theory, Einstein made the assertion that the speed of light is the same for all inertial observers, an inertial observer being one for whom the laws of physics take their simplest form. All observers moving with a constant velocity with respect to an inertial observer are also inertial observers themselves. If special relativity is indeed correct, and it has been proved to be so, then all laws of physics must be consistent with it. Einstein had found that the laws of electromagnetism were indeed consistent with special relativity. But the science of mechanics, which deals with the motion of bodies, and was based on Newton's laws of motion, was not consistent with the theory. So Einstein, in his special theory, modified what was known as classical mechanics, and in the process made the discovery that mass and energy were equivalent.

However, Einstein believed that his special theory was not the last word because of two reasons: (1) In special relativity, as the name implies, a special role is given to inertial observers. If someone is moving with acceleration, that is, with changing velocity, then that person's view of physics would be quite different from that of an inertial observer, which Einstein found to be unsatisfactory. (2) As mentioned in Sect. 3.3.2, special relativity was not compatible with Newton's theory of gravity. In Newton's theory, the gravitational force between one body and another propagates instantaneously, regardless of distance. On the other hand, special relativity requires that no body or signal can go faster than light. How is this contradiction to be removed? Einstein had successfully changed the laws of mechanics, but he was not able to similarly change Newton's theory of gravity to make it compatible with special relativity. Instead, he had to develop a completely new theory, which fundamentally changed our concept of space and time.

Einstein
Albert Einstein (1879–1955) was born in Germany. His father gave him a magnetic compass when he was a child. He was puzzled to see that the needle was always directed along one direction only. Since then he developed the habit of asking questions and searching for their answers. He was fond of mathematics from childhood and mastered algebra and geometry at the age of 12. After obtaining a diploma in a polytechnic in 1900 he failed to find a position as a teacher for 2 years. As a result, he took up a position as a clerk in the patent office in Berne, Switzerland. His job there was to examine applications filed for obtaining patents. There were several applications about methods for synchronising clocks which are far apart. These helped him in formulating his theory of relativity. He completed his Ph.D. in 1905. That year proved to be miraculous for him. During the year he published four papers which completely changed some of the fundamental ideas in physics. Each of these papers was worthy of a Nobel prize. Special theory of relativity was one of these. He was awarded the Nobel prize in 1922 for his work on the photoelectric effect. After 1908, he taught in several universities. In 1915, he stated his general theory of relativity which changed our concepts of space and time. This theory is considered to be the most creative and beautiful formulation by a human mind and Einstein is considered by some to be the greatest scientist ever. Einstein was offered the presidentship of Israel which he refused. He was an excellent violin player. He always upheld humanity and peace above everything else.

3.4.1   The Equivalence Principle

About 2400 years ago, the Greek philosopher and scientist Aristotle made the following assertion: if two bodies, with one being heavier than the other, were released from the same height, the heavier body would move faster and reach the ground first. About two thousand years later, Galileo Galilei performed experiments to show that Aristotle was not right in this regard. He demonstrated that all bodies, regardless of how heavy they are, always fall with the same speed when released from the same height. It is said that Galileo performed this experiment by dropping bodies from the famous Leaning Tower of Pisa, but the tale is probably apocryphal.

We now know that a body falls towards the Earth because of the gravitational force exerted on it. As a result of the force, the speed of the body continuously increases, and the body is said to be accelerated. Galileo's experiments show that all bodies, regardless of their mass, fall with the same acceleration. Near the surface of the Earth, this acceleration is about 9.8 metres per second squared (m/s^2). Galileo's conclusion is not limited just to the gravitational field of the Earth, it extends to the gravitational force exerted by all bodies. At any point in a gravitational field, all bodies, irrespective of their mass, would have the same acceleration. This is an expression of what is known as the *Equivalence Principle*. The principle was used by Albert Einstein in developing his general theory.

3.4.2   Formulation of General Relativity

In formulating general relativity, Einstein began, as was usual for him, from a very basic consideration. He used the principle of equivalence to argue that an observer with the right acceleration would not feel a gravitational field in the vicinity, even if a field were to be present. This can be understood by considering the experience that a passenger has in an elevator, or a lift as it is known in some countries. As the elevator starts moving towards a lower floor, the passenger feels somewhat lighter, as if there has been a reduction in weight. The opposite happens if the elevator starts moving towards an upper floor: now the passenger feels a sudden increase in weight. Einstein used this experience to create a famous thought experiment.

Consider an elevator like box suspended by a cable close to the surface of the Earth, with a person standing on its floor as shown in the left panel of Fig. 3.3. Now suppose the cable is cut, so that the elevator starts falling towards the Earth. Along with the elevator, the passenger in it will fall too and as we have seen above, the elevator and the passenger will both fall with the same acceleration, due to the equivalence principle. Now if the passenger drops a ball from his hand, the ball will not fall to the floor of the elevator as the ball and the elevator both are falling down with the same accelerated velocity and will be stationary with respect to each other. Thus, it will appear as if no gravitational force is acting on the ball. As the elevator is closed, the person inside will not realise that he is falling down and he will also not feel the force of gravity. Remember that the weight of a body is the gravitational force acting on the body due to the Earth. Thus, the result is that the passenger will be in a weightless state and will be able to float around the elevator, exactly like astronauts in a spaceship orbiting the Earth. This is shown on the right side of Fig. 3.3. The state of motion under the influence of gravity alone is called the state of free fall. In the above thought experiment, the elevator, the passenger and the ball are all in the state of free fall after the cable is cut. Einstein realised from this thought experiment that an observer in free fall in any gravitational field will not experience the gravitational field.

The converse of the above conclusion is also true. Consider an elevator in free space, far from all bodies. We can assume gravity to be absent there. Let us assume that there is a passenger in it who is weightless as there is no gravitational field. He will be floating freely, again like astronauts in a space ship. Now suppose the elevator is accelerated to one side at a constant rate, by a rocket attached to the opposite side of the elevator. While the elevator is accelerated, the freely floating person in it is not accelerated, as he is not touching the walls of the accelerating elevator. The result is that the person will land on the side on which the rocket is located. He will in fact be able to walk on it as if a gravitational field were present. If the acceleration of the elevator is 9.8 m/s^2, then the gravitational force that the person will feel would be the same as that on the Earth. Lesser acceleration would produce a weaker field and a greater acceleration would produce a stronger field.

From such thought experiments Einstein concluded that there is a similarity in the effects seen by an accelerated observer, and those seen in a gravitational field. This insight led Einstein to an immediate important consequence. Let us assume that another person standing outside an elevator is throwing a horizontal beam of light from a torch on the vertical wall of the elevator and that the beam is able to pass through the elevator. If the elevator and the passenger inside are stationary with respect to the torch, the passenger will see the beam of light to be travelling along horizontal direction as shown in the left panel of Fig. 3.4. If the elevator is going up with a uniform velocity, the passenger will see the beam to be travelling along an inclined straight line as shown in the middle panel of Fig. 3.4. If the elevator has an upward acceleration, then the passenger will see the beam of light to be travelling along a curved line as shown in the right panel of the figure. From this Einstein concluded that a ray of light in a gravitational field must follow a curved path. Light rays appear to travel in a straight line on the Earth as the gravitational field of the Earth is very weak and its effect on the path of light is too small to be noticed. Einstein was able to calculate the exact magnitude of deflection produced only after he developed the full theory, but the insight was very important for him in the development of the theory.

From the analogy between accelerated observers and gravitational fields, Einstein was able to conclude that the universal force of gravity was the manifestation of the bending of space and time. He postulated that gravity bends space and time and any curvature of the space-time is equivalent to the presence of a gravitational field. The higher the mass of an object, higher will be the gravitational field produced and larger will be the distortion of space-time. Another mass at a distance will feel this

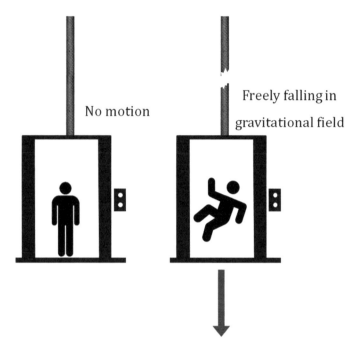

No motion

Freely falling in
gravitational field

Fig. 3.3 Einstein's thought experiment. On the left is shown a person in an elevator suspended above the surface of the Earth. On the right, the elevator is shown with the suspending rope snapped. In this case, the elevator falls freely towards the ground. The person no longer feels the gravitational force of the Earth. Any other objects in the elevator would float freely with the person, as explained in the text. Image credit: Kaushal Sharma

distortion even though no material medium is present. What does this mean? We can understand it with the following example.

Imagine a perfectly flat rubber sheet or mattress. If we place a heavy metal ball on it, the sheet will be depressed at the place under the ball, and also in the nearby region, with the depression decreasing with increasing distance from the ball. This is shown in Fig. 3.5. The heavier the ball, the larger will be the depression. Now imagine an ant walking on the mattress in the depressed region. The ant has decided to walk in a straight line from point A to point B and imagines it is doing so. The path of the ant is shown in the figure. Because the surface of the sheet is curved, in reality the path of the ant is not a straight line but a curved one in three dimensions.

The above example showed the distortion of a two-dimensional surface which is a rubber sheet. The space around us is three dimensional, and there is the fourth dimension of time. We know from special relativity that space and time are combined into a single four-dimensional entity called space-time. When a massive object like the the Sun is present in this space-time, it gets curved. It is difficult to imagine the four-dimensional space-time getting curved as we ourselves live in the

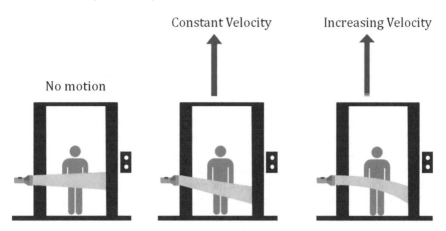

Fig. 3.4 Bending of light in accelerated frames. See the text for details. Image credit: Kaushal Sharma

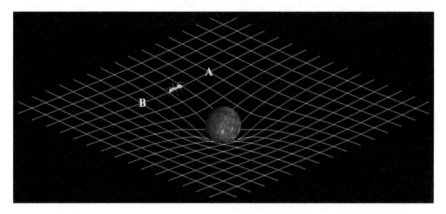

Fig. 3.5 The depression of a two-dimensional rubber sheet due to a heavy ball placed on it. Shown in the figure is the trajectory of an ant which believes it is moving in a straight line from point A to point B. Its actual path is a curved line in three dimensions, due to the depression in the sheet. Image credit: Kaushal Sharma

three-dimensional world with flowing time, but the curvature is mathematically similar to that of the rubber sheet.

Like the curved path of the ant in the above example, in the curved space-time around the Sun, a planet moves along a path which is known as a *geodesic*. This plays the role of a straight line in the curved space-time and is the path of minimum distance between two space-time points. When the geodesic is projected on three-dimensional space, the familiar ellipse-shaped orbit of a planet around the Sun is obtained. The curvature even affects the path of light rays which too in Einstein's theory move along curved paths in the presence of a gravitational field. This leads to interesting effects which are described below.

Even though there is no physical contact between the Sun and the planet and no medium is present between them (any gas present is extremely tenuous and there is near-perfect vacuum), the force of gravity of the Sun is conveyed to the planet by the curvature of the space-time itself. The effect is also not instantaneous but propagates forward with the speed of light. Thus, if the Sun suddenly lost half its mass, the effect would be felt on the Earth's orbit only after the time it takes for light to travel from the Sun to the Earth, which would be close to 500 s. It was shown by Einstein that his general theory is consistent with his special theory of relativity. So he did indeed succeed in creating a theory of gravitation which naturally took into account accelerated observers, was consistent with special relativity, and removed the difficulty of instantaneous propagation of the gravitational attraction between bodies, which was inherent to Newton's theory. Einstein thus solved the problems which he set out to, and in the process he created a completely new framework for physics. Space-time now became a dynamic entity, meaning that matter and, therefore, energy had an effect on its curvature. Matter and energy curve space-time, which in turn curves the paths of particles and light rays. The net effect is all the wonderful motion we see in the Universe, including the expansion of the Universe itself.

Einstein's theory of gravitation is fully described by a set of 10 complex equations which tell us how matter and energy generate gravitational fields through the curvature of space-time. There are further equations to describe the motion of particles and light rays in the gravitational fields. Finding solutions to these equations is, of course, very difficult, and over the past century, physicists and mathematicians have made serious efforts in solving them to find interesting new effects emerging from Einstein's theory.

3.5   Predictions of General Relativity and Their Observational Verification

The general theory of relativity is a new theory of gravity, offered as a successor to Newton's outstandingly successful theory. If general relativity is to be accepted as the correct theory, then (1) it should be able to clear basic, even philosophical, difficulties related to Newton's theory; (2) it should be able to explain all those phenomena like the motion of planets, occurrence of tides and so forth which were explained by the earlier theory and (3) it should make new predictions which can be tested and proven to be correct. We have already seen that the first requirement is amply fulfilled by general relativity. To verify the second point is straightforward. It can be shown that when gravitational fields are weak, and the speeds with which bodies move are small compared to the speed of light, then general relativity simply reduces to Newton's theory of gravity. The two conditions are fulfilled on the Earth and in the whole of the Solar system, so general relativity is indeed in agreement with Newton's theory. We are then left with examining whether general relativity makes testable new predictions to see whether it is a correct theory.

As soon as his theory of gravitation was completed, Einstein was able to show that there are three new results which the theory produces: (1) a precession in the orbit of the planet Mercury which exceeds the value predicted by Newton's theory; (2) a bending produced in the path of a light ray by the gravitational field of a massive body and (3) a change in the colour of light as it moves in a gravitational field. The last two effects are produced because light is also affected by gravitation like material particles are. This is explained in greater detail below. The three effects are all very small, and very accurate measurements have to be performed to verify them. This is in contrast to the situation with Newton's theory. When Newton developed his theory, it could immediately be used to precisely explain many observations which had accumulated over the centuries, like the motion of planets. The correctness and many applications of the theory were obvious to all scientists, and it was quickly accepted as the correct theory. Einstein's theory, on the other hand, was extremely difficult to understand, made few predictions and introduced altogether new concepts which challenged long held beliefs. People therefore were reluctant to accept the theory, until it was observationally shown to be correct, some years after the theory was first presented in 1915. We will now describe the three results mentioned above.

3.5.1 Precession of the Orbit of Mercury

As we saw in Sect. 3.3.2, the elliptical orbit of the planet Mercury rotates in space at a slow rate. There is a small excess of 43 arcseconds per century of the observed value of this precession over the value predicted by Newton's theory. Einstein showed that the excess is exactly matched by the prediction made by general relativity for the precession. This was a great triumph of the theory, but it did not fully capture the attention of the public and that had to wait until 1919.

3.5.2 Gravitational Bending of Light

Light was known to travel in a straight line as proved in numerous experiments on the Earth, and also by the occurrence of eclipses which form due to the shadow of the Earth and Moon; the phenomenon of shadow works due to the straight line propagation of light. According to Einstein's theory, in the absence of a gravitational field, a ray of light would travel in a straight line. However, if the space is warped due to the gravity of a massive object, light will be forced to travel a curved path. Light will thus no longer travel in a straight line but its path will bend while passing close to a massive object.

Another way to understand the bending of light is as follows. As energy and mass are equivalent (see Sect. 2.4.5), light, which has energy, can be considered to have an equivalent mass. It will, therefore, feel the gravitational force and its path will deviate from being a straight line. The masses of objects on the Earth are too small

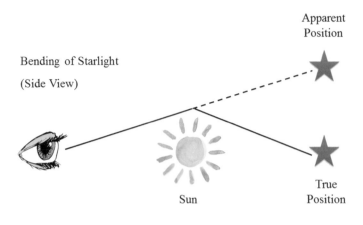

Scale is exaggerated

Fig. 3.6 The bending of star light due to the gravity of the Sun. The observer sees the position of the star shifted relative to its true position. The actual change in the direction of the ray is gradual and much smaller than shown. Image credit: Kaushal Sharma

to cause detectable bending of light which therefore remained undiscovered. But the Sun has a large mass and a light ray from a source behind it, passing close to the Sun, would bend by an angle which can be measured on the Earth. This is shown in Fig. 3.6. It is seen from the figure that a star behind the Sun would appear to be displaced because of bending of the rays coming from it to us.

The light we receive during the daytime from the Sun is very bright. Therefore, the light from any stars close to the Sun, or the bent light from a star behind it, would simply not be visible against the bright background. But there are very special occasions, when the Sun is up in the sky, and yet the light from it is not received on the Earth. These are, of course, total Solar eclipses during which the light from the Sun is completely blocked by the Moon and it gets dark on the Earth. At such a time, we can see stars in the sky. During a total Solar eclipse, if there is a star close to the Sun's disc, and rays of light from it are bent from their path due to the gravity of the Sun, then the star will appear to be displaced from its actual position to a new apparent position. We can determine the apparent direction of such a star during a total Solar eclipse and determine the real direction at other times, say 6 months later when the Sun is not close to it and when we can see it at night. The difference between the directions gives the amount of bending produced, and this can be compared with the prediction of general relativity, which is 1.75 arcsecond.

An opportunity of measuring the bending of light presented itself on the 29 May 1919, 4 years after the birth of the theory, when a total Solar eclipse was to be seen in parts of South America and Africa. The famous British astronomer Sir Arthur Eddington and his colleagues organised expeditions to measure the position of stars near the Sun during the total eclipse. Eddington led a team to the island of Sao Tome off the coast of Gabon in Africa, while another team went to Sobral in Brazil. From analysis of data obtained during the eclipse, the positions of the stars were indeed found to have shifted from their normal positions, in accordance with predictions of the general theory and proved that light bends due to the gravity of a massive object. Einstein soon became a household name as the scientist who had disproved Newton, the greatest scientist known till then.

A sidelight to this eclipse story is that in 1911, using the equivalence principle described in Sect. 3.4.1, Einstein had predicted that light should bend from its straight path in a gravitational field. He calculated a value of 0.87 arcsecond for the bending of light produced by the Sun, which is half the value predicted by general relativity. Attempts to observationally test Einstein's early prediction were unsuccessful for various reasons, which was in fact fortuitous! Successful observation would have shown that the observed value of bending was twice that predicted by Einstein on the basis of the equivalence principle, which would have been a great set back for Einstein's ideas.

Gravitational Lensing: The gravitational bending of light, which appears as a very small deviation from a straight path in the Solar system, can lead to many interesting effects, which are broadly known as *gravitational lensing*. Einstein considered a situation when a star or galaxy which is far away, which we will call A, lies exactly behind a nearer star or galaxy, which we will call B. In such a case, most of the light rays emitted by A, if they travelled in straight lines, would have passed us by; those which were travelling along the line AB, would have hit B and either would have been absorbed there or would have been scattered and we would not have been able to see A. However, as per Einstein's theory, the divergent rays from A, which pass close to B will be bent from their path due to its gravity and will be able to reach us. This is shown in Fig. 3.7. Thus, one can see light from A coming to us from all directions along the cone centred at the Earth as shown in the figure. This means that we will see a ring instead of a point source like a star. This is called *Einstein's ring*.

As the stars and galaxies are very thinly distributed in space, the probability of two sources being exactly aligned is extremely small. Also, the radius of the ring would be so small that even if it were to be observed, it would have appeared as a point source, rather than as a ring with the then available telescopes and instruments. Because of these reasons, Einstein dismissed the idea that such an occurrence could ever be seen. If the alignment of A and B is slightly off from being exact, i.e. A, B and the Earth do not lie on a straight line, the symmetry gets broken and so if B is slightly away from the line joining us and A, in place of a ring, we would see multiple images of A. As glass lenses are used to bend light and form images in laboratories, the above-mentioned phenomenon is called gravitational lensing. As in the case of more familiar optical convex lenses, gravitational lenses too converge light rays and so they make an image brighter than it would have been in absence of lensing. This can help us to observe distant objects which would otherwise be too faint to observe.

With the progress in technology, in 1979, three astronomers, Dennis Walsh, Robert Carsewell and Ray Weyman, actually observed a twin image of a very distant source, called a *quasar*. Quasars are intrinsically very bright and thus can be observed to very large distances. The astronomers observed two quasars which appeared to be

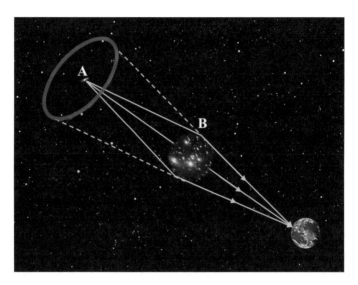

Fig. 3.7 A sketch of the ring visible to an observer on the Earth who is exactly in line with two astronomical sources. The ring is the image of the background source A made because of bending of rays by the gravity of the foreground source B. Image credit: Kaushal Sharma

very closely spaced in the sky. They had identical properties and a detailed study confirmed that these two quasars were not different but were two images of the same quasar, gravitationally lensed by an intervening galaxy. Since then a large number of multiply imaged sources (i.e. gravitational lenses) have been observed. Einstein's rings have also been seen. Six examples of multiple images are shown in Fig. 3.8 and an example of Einstein's ring is shown in Fig. 3.9. Gravitational lensing is now a very important tool for astronomers to study various aspects of the Universe.

3.5.3 Gravitational Redshift

We know that a ball thrown vertically from the surface of the Earth gradually slows down, eventually stops and returns to the Earth with increasing speed (unless it is thrown with a speed exceeding 11.4 km/s, in which case it escapes from the Earth). The ball slows down and returns to the Earth because of the force of gravity acting on it. We have seen that in general relativity material particles as well as light are affected by the gravitational field. Now imagine that a torch is shone pointing upwards from the ground. The rays of light travel upwards and these too are affected by the gravitational field. They cannot slow down as the ball does, because light must always travel at the same constant speed, which is 300,000 km/s in vacuum and slightly smaller than this in the Earth's atmosphere. The light ray, however, will lose energy, exactly as a ball that is slowing down does. The effect then is to increase

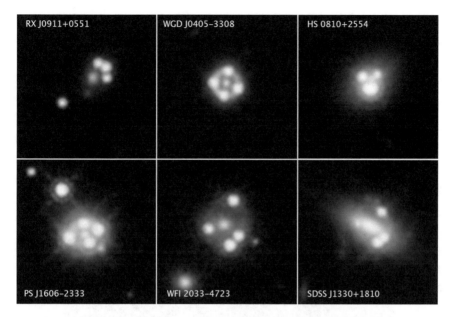

Fig. 3.8 Each of the six frames, taken by the Hubble Space Telescope, shows four distorted images of a distant quasar formed due to gravitational lensing by a galaxy. In each case, the lensing galaxy, which is seen in frame is along the line of sight in the foreground, i.e. the galaxy is much closer to us than the quasar. Image credit: STScI

the wavelength of the light, that is, the light becomes redder. This is known as the *gravitational redshift* of light. Similarly, if a torch was shone towards the ground from a height, the ray gains energy, just as a ball speeding up on its way downward does. The effect now is to make the light bluer in colour, due to a decrease in its wavelength, which is known as the gravitational blueshift of light.

It was difficult to experimentally verify this prediction because the effect is very small in the Earth's gravitational field, and the experimental accuracy available in the early part of the twentieth century was not enough for the purpose. But help with this matter came from *white dwarf* stars. We will discuss these stars in detail in Chap. 5, here it is enough to know that they have a mass comparable to that of the Sun but have a radius which is about a hundred times smaller. The gravitational field at their surface is, therefore, very much stronger than the field at the surface of the Earth or the Sun. Using the spectrum of the white dwarf Sirius B, W. S. Adams in 1925 reported that absorption lines in the spectrum of Sirius B are indeed redshifted. While there are some doubts now about the correctness of the measurements obtained by Adams, modern observations have provided independent evidence of redshift in the light from Sirius B. Gravitational redshift was first measured on the Earth in 1959 by Robert Pound and his student Glen A. Rebka using the Mössbauer effect.

The change in the colour, that is, the change in wavelength, of light occurs in general relativity because the rate at which clocks run depends on the intensity of the

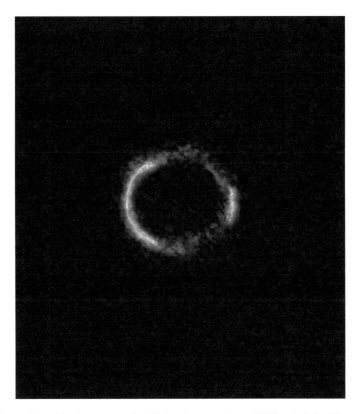

Fig. 3.9 An Einstein's ring observed with the Atacama Large Millimeter/submillimeter Array (ALMA) radio telescope. A distant galaxy SDP.81 and a foreground galaxy line up so perfectly as seen from an observer on the Earth that the light from the distant galaxy forms a nearly perfect circle due to gravitational lensing. The foreground galaxy which acts as a gravitational lens is too faint to be observed. Image credit: ALMA (NRAO/ESO/NAOJ); B. Saxton (NRAO/AUI/NSF)

gravitational field. If one clock is located on the surface of the Earth and another at a great height, where the gravitational field of the Earth is small (the field decreases with increase in height above the Earth's surface) then it can be shown that the clock high above runs faster than the clock on the ground. The time difference is small and can be counted in nanoseconds (1 nanosecond is one billionth of a second, that is, 10^{-9} s). While this may appear to be inconsequentially small, in these modern times it does affect us at the practical level. For example, it is essential to take the slowing down of clocks into account for proper functioning of the global positioning system (GPS). The functioning of this system depends on the time measured by clocks which are carried by satellites. The satellites are in orbit around the Earth at great heights, where the gravitational field is smaller than on the ground. So the clocks in the satellites run faster than clocks on the ground. The fact that the satellites move at high speed also affects the rate of the clocks due to special relativity. We are very dependent on GPS systems in several ways. They are used in communication and are indispensable for navigation in civil as well as military situations. Many of us even use GPS in cars. Google Maps use GPS. If we used Newton's theory in the GPS instead of Einstein's theory, our positioning would be wrong by about 8 km per day. So an effect which is very small, and which at one time could be detected only with great difficulty, is now important for us in our day-to-day lives.

Star S2

In April 2019, astronomers have reported to have measured the gravitational redshift of absorption lines produced by hydrogen and by helium in the spectrum of a star named S2, which is orbiting very close to the centre of our galaxy. The Milky Way harbours a *supermassive black hole* at its centre. The mass of this black hole has been estimated to be about 4×10^6 times the mass of the Sun by studying the motion of star S2 which orbits the black hole at the centre of the Galaxy. The lines of hydrogen and helium from S2 are redshifted due to the strong gravity of the black hole. As S2 moves through the changing gravitational field of the black hole, the gravitational redshift changes. Measurements show that the gravitational redshifts of both lines are identical as expected in Einstein's theory. This provides proof of the theory under the strong gravitational field of the supermassive black hole.

3.5.4   Beyond the Classical Tests

The three tests of general relativity discussed above are known as the classical tests, because they were discussed as the first tests of the theory. Over the years, the measurements of light bending and of the gravitational red and blueshifts of light have improved with advances in technology. Increasing accuracy has been obtained from tests performed on the Earth, as well as using clocks and detectors placed on aeroplanes and spacecraft. Predictions made by the theory beyond the three tests are also being tested through experiments. The Apollo astronauts left reflectors on the Moon, which have been used in Lunar laser ranging experiments. In these, a laser beam is sent from the Earth to the Moon and is reflected back to Earth. The time taken by the beam to travel forth and back is measured accurately, which leads to a very accurate measurement of the distance to the Moon. Lunar laser ranging has also been used to perform tests of general relativity. In every test made so far, the predictions of general relativity have been confirmed to a high degree of accuracy. Einstein's theory has remained the best theory of gravity so far, performing better than many other theories which have been proposed as replacements for Newton's theory of gravity.

In 1916, Einstein used his theory to predict the existence of gravitational waves, like the prediction of electromagnetic waves from Maxwell's theory of electromagnetism. It took many years for even the theoretical analysis related to gravitational waves to be established as being correct. Gravitational waves will be discussed in the next chapter and their detection will be discussed in detail in later chapters.

Chapter 4
Gravitational Waves: The New Window to the Universe

Abstract In this chapter, we describe gravitational waves. These were first predicted by Einstein in 1916, using his new theory of gravitation which he had announced a year earlier. Einstein himself did not fully believe his own result, and it took decades of theoretical work before gravitational waves were accepted as real. While electromagnetic waves are space- and time-dependent undulations of electric and magnetic fields, gravitational waves are undulations in the fabric of space-time. Both travel with the speed of light and can be polarised. Electromagnetic waves can be easily produced by sources in the laboratory and can be equally easily detected. Gravitational waves interact very weakly with matter, and so they are difficult to detect. Presently detectable gravitational waves are only those produced by very massive cosmic sources with large accelerations. Such waves travel unhindered through space and produce only tiny effects on the Earth.

4.1 Introduction

In this chapter, we will describe gravitational waves, which are the main subject matter of this book. In Chap. 2 we discussed different waves like the water waves, waves on a string or rope, sound waves and finally the electromagnetic waves. Gravitational waves are of an entirely different type. They were predicted by Einstein in 1916, a hundred years before their experimental discovery. There was much debate among the scientists about the correctness of the mathematics that led to their prediction, and the reality of the presence of such waves was not accepted immediately. In fact, in 1936, Einstein himself claimed that gravitational waves cannot exist, but changed his ideas after peers pointed out the fallacy in his arguments. Even though a number of scientists contributed to the final acceptance of gravitational waves as a natural outcome of Einstein's theory of general relativity, Einstein is credited with their prediction. An indirect observational proof of their existence was obtained only in 1974, some 20 years after Einstein's death. We will discuss that in the next chapter. Their direct detection, however, had to wait till 2016 as we will see in detail in later chapters. Below, we will describe these waves and then look at their possible sources.

4.2 The Metric

As seen in Sect. 2.3, some quantity increases and decreases periodically in every wave. In the water waves, it is the height of water particles, from their mean position of rest, which changes with time and with distance as the wave propagates. In electromagnetic waves, the magnitudes of the electric and magnetic fields change with time and distance. In the case of gravitational waves, the situation is more complicated, because the change over time and distance, i.e. space, must occur in the gravitational field. But in Einstein's theory, the gravitational field itself is at the basis of the definition of space-time. It is therefore difficult to distinguish between real changes in the gravitational field from changes merely in the way space and time are measured, and that is the reason why it took so long for scientists to accept the reality of gravitational waves after their first prediction in 1916.

We have seen in Chap. 3 that in Einstein's general theory of relativity, the force of gravity is explained as being due to the *curvature of space-time*. The presence of matter and energy curves space-time, which affects the way particles and light rays move. The greater the curvature, the greater is the gravitational force. The curvature of space-time is the main entity in the general theory.

The measure of curvature is provided by mathematical quantities which are known as the Riemann and Ricci tensors, which can be calculated from a much simpler mathematical quantity which is known as the metric tensor, or simply the metric. The geometry of a circle or of a spherical surface like the surface of the Earth is described completely by the radius of the circle or sphere. Similarly, the geometry of any space, i.e. its curvature at any point of the space, is fully determined if its metric tensor is known at all points. The space could be the usual three-dimensional space that we are used to in our everyday experience, or any other space with more complicated properties. This also extends to the four-dimensional space-time of special theory and general theory of relativity. In the special theory, three dimensions of space and time are combined into a single four-dimensional entity. The form of the metric for this space-time is very simple, with the details depending only on the way in which space and time are measured. The space-time here is said to be flat. In the general theory, as we have already described in Sect. 3.4, the geometry is a dynamic entity. The mathematical form of the metric therefore depends upon the distribution of matter and energy which are present. Given these distributions, *Einstein's equations* can be solved to obtain the metric. It is in general very difficult to find solutions to these equations, and only a small number of useful ones are known, like the *Schwarzschild solution* we have mentioned in the context of black holes in Sect. 5.4.3.1. The difficult task of solving the equations can become somewhat simpler when the gravitational field is weak. In this case, the metric can be expressed as a sum of two parts, one the flat metric of special relativity which is already known, and another a relatively smaller part describing the weak gravitational field. It turns out that the weak field satisfies a simpler set of equations which can be more easily solved. Einstein used this method, known as the weak field approximation, to make predictions about gravitational effects in the Solar system, including the bending of light before the first exact solution was found by Karl Schwarzschild in 1916.

4.3 Weak Fields and Gravitational Waves

The Sun is a spherical object with a large mass and radius, which remains unchanged over very long periods of time. The gravitational field of the Sun, therefore, remains *static* or unchanging with time. As a result, solutions of Einstein's equations which apply to the Solar system, like the Schwarzschild solution, or the weak field approximation solution mentioned above, are not time dependent and are relatively easy to obtain. But the situation is quite different when we consider solutions which can correspond to gravitational waves. As a wave propagates, the amplitude changes with time as well as space. So we have to necessarily solve equations which are dependent on time, which is much more difficult to do than in the static case.

In the time-dependent case, the metric is again written as the sum of the constant, flat metric of special relativity, and the weak field part which is now space and time dependent. When a certain condition, known as the Lorentz gauge condition, is applied to the weak part of the metric, it can be shown that the weak field satisfies the wave equation. This is a very important result, as it shows that the gravitational field propagates as waves, at least when the field is weak and far from all sources of gravitation. The periodic changes in the metric which we are interpreting as gravitational waves have another connotation of course. The metric determines the geometry of space-time, and therefore these waves are often described as *ripples in the fabric of space-time*. We may draw an analogy here with the waves on a two-dimensional fabric surface when two women hold the ends of long piece of fabric and move it up and down creating a wave, as shown in Fig. 4.1. It follows from the wave equation that the speed of propagation of the gravitational waves is just the same as the speed of propagation of electromagnetic waves, which is equal to the speed of light.

Fig. 4.1 Waves on a two-dimensional fabric which is being aired by two women. Image credit: Kaushal Sharma

4.4   Gravitational and Electromagnetic Waves: Similarities and Differences

There are several similarities and differences between the electromagnetic and the gravitational waves as seen below.

4.4.1   Similarities

As described in Sect. 2.3.1, once Maxwell formulated the equations of electromagnetic theory, he was able to show from the equations that the electric and magnetic fields satisfy the wave equation when there are no electric charges or currents present. Similarly, in the case of gravitation, the wave equation for the metric follows from Einstein's equations in absence of sources. Other similarities are that both gravitational waves and electromagnetic waves travel with the velocity of light in vacuum. Also, gravitational waves, like electromagnetic waves, can undergo redshift and can be gravitationally lensed, and are *polarised*, that is the vibration of the field is in specific planes.

4.4.2   Differences

A major difference between gravitational waves and electromagnetic waves is that the gravitational wave equation applies only when the gravitational field is weak, while the wave equation for the electromagnetic field applies even when the field strength is arbitrarily large. Another major difference is that electromagnetic waves are affected by matter and therefore cannot propagate over long distances in the presence of matter without getting significantly diminished in intensity. Gravitational waves, on the other hand, are unaffected by the presence of matter and can travel long distances. This is mainly due to the fact that the gravitational force is extremely weak, being the weakest of the four fundamental forces, while the electromagnetic force is by far stronger, being the second strongest of these forces. Electromagnetic waves therefore easily interact with any electric charges that may be present along the path, and are either absorbed or scattered into another direction. The fact that gravitational waves do not significantly interact with any matter that may be along their path is very important because, using gravitational waves, we can observe far away sources in extreme conditions, and even in the conditions which existed in the very early epochs of the Universe.

The second difference between the two is that the wavelengths of electromagnetic waves are typically smaller than the size of the cosmic sources in which they arise, while it is the opposite for gravitational waves. Thus, the wavelengths of gravitational waves are larger than a few km and can be as large as 10^{21} km, while those of electromagnetic waves are always smaller than a few km and can be smaller than the size of an atomic nucleus as seen in Sect. 2.3.2.

The third difference is that the gravitational interaction being extremely weak, it is very difficult to detect gravitational waves, while the electromagnetic interaction is much stronger and hence the electromagnetic waves are much easier to detect.

The fourth difference, about which we will not go in details but simply mention here for completeness is that the electromagnetic waves are mainly dipolar in nature while the gravitational waves are mainly quadrupolar in nature.

4.5   Emission of Gravitational Radiation

We have seen in Sect. 2.2.3 that electric charges and currents are the sources of electric and magnetic fields and that accelerated charges produce electromagnetic radiation. Similarly, masses are the sources of gravitational fields and accelerated masses generate gravitational waves. Thus, even when we move a finger gravitational waves are produced. However, gravitational waves carrying significant energy are only produced by very large and highly accelerated masses, with certain conditions satisfied, which we will describe below.

The nature of the sources which can emit gravitational radiation follows from a detailed analysis of the gravitational wave equation. It can be shown, for example, that an expanding or contracting uniform sphere of matter, accelerating during the expansion and contraction, will not emit gravitational waves, because of the spherical

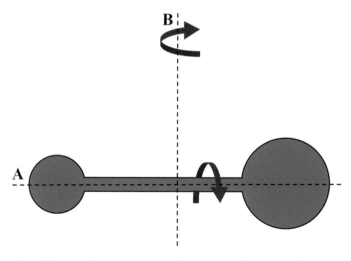

Fig. 4.2 A dumbbell which is symmetric around the axis A and is asymmetric around axis B will not emit gravitational waves if it rotates around axis A but will emit these waves while rotating around axis B. Image credit: Kaushal Sharma

symmetry. Similarly, a cylinder or a disc of matter or a dumbbell which is rotating around its axis of symmetry cannot emit gravitational radiation because of the cylindrical symmetry. But the same dumbbell rotating around an axis perpendicular to its length can emit gravitational radiation as shown in Fig. 4.2. The emitted gravitational radiation has different components, the most important of which is called quadrupole radiation. The other components present are much weaker than the quadrupole part.

The objects that we come across in our day-to-day life all have very small masses and hence the gravitational fields of these objects, and any gravitational waves generated by them, are very weak. By weak here we mean that the magnitude of the gravitational field in the waves, which varies with space and time, is very small. Even the gravitational waves generated by planets and stars like the Sun are very weak. Thus, the displacement of a particle under the influence of these gravitational waves will be too small to be detected.

We will discuss about the indirect and direct detection of gravitational waves in more detail in later chapters. Here we note that it is beyond our present capability to construct a machine on Earth which will be able to generate detectable gravitational waves. We thus have to turn to much larger masses and much higher accelerations than are found on the Earth or found even in the Solar system. In the next chapter, we will discuss the possible astronomical sources which may produce gravitational waves which we can hope to detect.

Chapter 5
Compact Sources of Gravitational Waves

Abstract In this chapter, we describe how compact cosmic objects like white dwarfs, neutron stars and black holes are formed. Stars for much of their life produce energy, through nuclear fusion, which is emitted from their surface. In this stage, stars have a fixed size which is maintained by balancing their own gravitational field with the high pressure of their gas. Once the fuel for the fusion is exhausted, the central region of the star collapses to a compact object, with the rest of the star exploding into space. The compact object can be either a white dwarf, neutron star or black hole depending on the mass of the original star. Some of the neutron stars can be detected as radio pulsars, which emit narrow beams of radio and other radiation. Understanding these compact objects requires a combination quantum mechanics, the special theory of relativity and the general theory of relativity. It is the massive and very compact neutron stars and black holes which are of interest to us in the context of gravitational waves. When present in compact binary systems, they undergo large accelerations, emitting gravitational waves which can be detected. Gravitational waves can also be produced by asymmetric core-collapse supernova explosions, Gamma ray bursts, pulsars with deformities and as a stochastic background.

5.1 Introduction

As we saw in the last chapter, gravitational waves produced by most objects on the Earth or even in the Solar system are very weak, in the sense that the effects they produce are too small to be detected even by the most sensitive instruments on Earth. For gravitational waves to be detectable, the gravitational field that they carry to the Earth should be sufficiently strong to be able to make changes in a detector which can be interpreted as gravitational waves. For that the waves have to be produced by objects which are very massive and compact, so that their gravitational field is very strong. Moreover, the objects have to undergo large acceleration, so that they can radiate strong gravitational waves. Such objects are rare and they are only found in distant regions of the our galaxy and in other galaxies. Gravitational waves emitted by these sources have to travel over a great distance to reach the Earth, so the energy is spread over a very large sphere whose radius is equal to the distance between the emitter and the Earth. Therefore, however strong the gravitational waves at the sources may be, we receive only a very weak flux at the Earth, which makes the detection very difficult.

Most astronomical sources have large mass, but not all of these sources undergo significant acceleration. The highest accelerations are obtained by either the material of exploding stars or by the motion of compact stellar remnants in binary systems. In this chapter, we will describe how stars evolve, how they explode towards the end of their life cycle, the nature of the compact remnants that they leave behind and how such remnants, when located in binary systems, are strong emitters of gravitational waves. Our understanding of these matters is based on theoretical modelling as well as observations. Objects like the stars are not easy to model as there are so many different processes which take place inside them. However, with four centuries of observations, more than a century of theoretical work and more than half a century of work using computers, we have a lot of confidence in our present understanding.

5.2   Structure of Stars

A star is a very large sphere of gas which emits a huge amount of energy. There are about 4000 stars in the sky visible at night to the naked eye. While most stars appear to be yellowish in colour, it is easy to see that there are stars with other colours as well, like red or blue. Unlike planets, whose positions in the sky keep changing on a day-to-day basis, stars seem to remain at fixed points in the sky. But it was discovered even in ancient times that stars do move a little, when observed over many years. This led to the idea that stars were much further away than the planets. We now know that the stars we see with our eyes in the night sky are a very tiny fraction of about a hundred billion stars that constitute the Milky Way galaxy. These stars are spread in a disc-like structure which has a diameter of about a hundred thousand light years.

Over the last hundred years or so, using large telescopes and sophisticated instruments like spectrographs, astronomers have been able to measure distances to a number of stars, their mass, size, temperature, chemical composition and other properties. Applying the laws of physics to this data, astronomers have developed a very good, but not yet complete, understanding of the nature of stars, how they are formed, how they generate energy, how long they will live and what happens to them at the end of their life. We will provide a brief description of these matters in the present section.

The Sun is the star nearest to us and we depend on it for our existence and sustenance. The Sun is a huge sphere of hot gas with a mass of 2×10^{30} kg, a radius of 7×10^5 km and is at a distance of 150 million km from the Earth. The yellowish colour of the Sun tells us that the temperature at the surface of the Sun is about 5800 K (see Sect. 2.5). About 72% of the mass of the Sun is made up of hydrogen gas, about 26% of helium and the rest consists of heavier elements like carbon, nitrogen, oxygen, etc., all in gaseous form.

The Sun is a typical or average star in our Galaxy, in the sense that there are (i) many stars which are more as well as less massive than the Sun, (ii) many stars which are hotter as well as cooler than the Sun and (iii) many stars which are bigger as well as smaller in size as compared to the Sun. We can therefore use the properties of the Sun as a unit: the masses of stars are stated as a multiple of the Solar mass and the radii of stars are stated as a multiple of the radius of the Sun. In these Solar units, stars have mass in the range of about 0.1 times to a few hundred times the Solar mass, and the radii of stars are in the range of a fraction of the Solar radius to about a 1000 times the Solar radius. The present age of the Sun is about 4.5 billion years, and we expect from the theory described below that it has a total lifetime of about 9 billion years. For most stars, the lifetime varies from a few million to ten billion years or more. The higher the mass of a star, the shorter is its lifetime.

5.2.1 Hydrostatic Equilibrium

The Sun has had almost the same properties, like temperature, mass and size, as it has now, for most of the past 4.5 billion years. As per our current understanding of stars, these properties of the Sun will remain almost the same in the future for a similar duration. Why is it that the Sun will remain in the same state for nearly 9 billion years? The reason is that it is in a near-perfect *mechanical equilibrium* state. There are two competing forces that have opposite effects on the matter in a star. One is the familiar attractive force of gravity which tries to pull all the matter of the star together to one point, its centre. Due to the large mass of the star, this force is so strong that if it were unopposed, the star would collapse to a point in a couple of hours. Obviously, for a star to have remained of the same size for billions of years, the force of gravity needs to be opposed and balanced exactly by another force. This opposing force is provided by the *pressure* of hot gas.

We know that gas exerts pressure from the fact that when gas is inserted in a balloon, the balloon expands. The sides of the balloon are pushed outwards due to the pressure of the inserted gas. The hotter the gas, the higher is the pressure it exerts. Also, the higher the density of the gas, the higher is the pressure. This dependence of gas pressure on the temperature and density is true for ordinary gas all around us and elsewhere in the Universe. Departure from this dependence occurs in extreme conditions which occur inside compact stars as we will see later in this chapter. The hot gas comprising the star tries to disperse into the near-vacuum conditions of the surrounding space and thus provides an outward force, opposing gravity. For stars like the Sun, both these opposing forces exactly balance each other. This is known as *hydrostatic equilibrium*.

The hydrostatic equilibrium of the star means that the pressure at any depth inside the star needs to be high enough to support the weight of all the matter above it. As we go deeper, the weight to be supported is greater, so the pressure is higher. This in turn implies that the density and the temperature must increase as we go closer to the centre. Using the equations which describe the hydrostatic equilibrium, it can be shown that the temperature at the centre of a star having the mass of the Sun is about 20 million K, which is much higher than the temperature at the surface. The larger the mass of a star, the higher is its central temperature as a larger weight has to be supported by the pressure at the centre. This needs higher temperatures. Hence, the central temperature is proportional to the mass of the star, a fact which is crucial to understanding the evolution of the star over its lifetime.

5.2.2 Generation of Nuclear Energy

The Sun, being much hotter than the surrounding cold space, is constantly losing heat to it. We know this very well as all the energy necessary for our survival, in the form of radiation and heat, comes from the Sun. The energy lost must come from the interior of the Sun as the temperature towards the centre is much higher than the temperature at the surface. Due to this flow of heat from the inside to the outside, the Sun would begin to cool, unless there is generation of energy which balances the outflow of heat. The force of gravity in the Sun does not change with time, because its mass and size do not change. So for the gas pressure to keep balancing it, as it seems to be doing, the temperature has to be maintained. It is critical that there is some source of energy in the Sun which will maintain the temperature.

The source of energy is the nuclear energy generated due to the fusion of hydrogen, which is the most abundant element in the Sun, to produce helium. This process requires very high temperature and density which are found only at the centre of the Sun. Thus, the nuclear energy generation occurs only in the central region. The conversion of hydrogen to helium with the release of energy happens in all stars which are at the same stage of their evolution as the Sun.

5.3 Evolution of Stars

The balance between the inward force of gravity and the outward pressure is disturbed once the hydrogen in the central region of the star, and hence the supply of energy, is exhausted. This is what will happen to the Sun after about 4.5 billion years. In absence of a source of energy, the temperature starts falling as the star continuously keeps emitting stored heat energy on account of it being hotter than the surroundings. As seen above, a decrease of temperature causes gas pressure to decrease and it can no longer balance the gravitational force of attraction. This leads to a contraction of the inner regions of the star. This is similar to what will happen to an inflated balloon if it is placed inside a refrigerator; it will begin to shrink because the pressure of the gas inside it decreases with decrease in temperature. Because of contraction, the *gravitational potential energy* of the star decreases, which results in an increase in the thermal energy of the star, as the total energy has to always remain the same as per the laws of physics. Thus, the contraction of the star heats up its gas. The density of the matter in the star obviously increases as the star contracts and becomes smaller in volume while its mass remains the same.

With increased density and temperature, the conditions at the centre become right for the start of the next stage of nuclear energy generation. In the first stage, hydrogen was converted to heavier element helium. In the next stage, helium, in turn, is converted to even heavier elements like carbon and oxygen, in the process generating energy. Why does helium not undergo fusion at the same time when hydrogen does? That is because, the heavier an element is, the higher is the density and temperature required for it to undergo nuclear fusion. That is in turn because, to undergo fusion, two nuclei of that element have to come extremely close to one another so that they can experience the nuclear, i.e. the strong force which is responsible for fusion. The nuclei are positively charged. A hydrogen nucleus has the smallest positive charge as it has only one proton, while helium nuclei have two protons and two neutrons. Because the latter are particles without electric charge, a helium nucleus has twice as much positive charge as a hydrogen nucleus. As a result, the force of repulsion between two helium nuclei is four times as high as that between two hydrogen nuclei. To overcome this larger force, in a collision the velocities of helium nuclei should be much higher than those needed for hydrogen nuclei. As velocities are directly proportional to thermal energy, the higher velocities mean higher thermal energy, i.e. higher temperature. The fusion of helium requires a temperature of about 100 million K, which is very much higher than the temperature at which the fusion of hydrogen takes place. So the conversion of helium begins only after the central region has contracted enough for the helium there to reach the very high temperature required. With this new source of energy, the star again achieves equilibrium. Such cycles of one type of fuel getting exhausted leading to contraction of the star, its getting hotter and denser, and heavier elements undergoing fusion and supplying energy and imparting temporary equilibrium to the star continue to occur one after another. How many such cycles take place depends on the mass of the star as we will see below.

The amount of nuclear fuel available to a star is proportional to its mass, while the rate at which this fuel is expended is proportional to the luminosity of the star, which is the amount of energy released by the star per second. The luminosity increases very rapidly as the mass of the star increases, the net result being that the more massive a star, the shorter is its total lifetime. A star which has a mass 10 times the mass of the Sun would have a lifetime of only about one thousandth of the lifetime of the Sun. Since the Sun will live for about 9 billion years, a star which has 10 times the mass will live only for 9 million years, which is a rather short time for astronomy.

5.4 End Stages of Stars

We will now look at further evolution of stars and the end stages of their life. The evolution of a star beyond hydrogen fusion depends critically on the original mass of the star and less importantly on its chemical composition. In order of increasing mass, stars end their lives as white dwarfs, *neutron stars* or black holes as described below. The first two of these objects have mass comparable to the mass of the Sun, but a radius which is very much smaller, so that their density is very large and the gravitational field at their surface is very strong. Black holes which are produced as a result of stellar evolution can have much larger masses and their radius is zero. The physics of black holes is therefore quite different from the physics of the other compact objects. Let us look at the evolution of stars with different mass ranges.

5.4.1 Stars with Mass Less Than Eight Solar Masses

Objects which have mass smaller than about 0.08 times the mass of the Sun do not ever become stars in the sense that nuclear energy generation never can start at their centres. The reason for this is easy to understand. We have seen in Sect. 5.2.1 while considering hydrostatic equilibrium that the temperature at the centre of a star depends on its mass. When the mass of a star is less than about 0.08 times the Solar mass, the highest temperature achieved at the centre is smaller than 2 million K. At these relatively low temperatures, hydrogen cannot undergo nuclear fusion. These objects thus remain as dark objects which are called *brown dwarfs*, firstly as they are small in comparison to most stars and secondly, because they emit very little radiation. This radiation is due to the heat generated in these objects, because of the contraction they undergo at birth and is mostly in the infrared part of the spectrum. Ultimately, after much of their heat energy is lost, the brown dwarfs become *black dwarfs*. If the mass of the object is less than about 0.01 Solar masses to begin with, then it becomes a planet, rather than a brown dwarf.

Stars which have mass above the limit for brown dwarfs, and below about eight Solar masses, go through at most three of the cycles of nuclear fusion mentioned above. The lower mass stars in this range will stop after the first cycle, i.e. after hydrogen is converted to helium. The reason again is that because of their low mass, upon contraction after the end of hydrogen fusion, they cannot reach the high temperature which is needed for the fusion of helium. The somewhat more massive stars may go through the second cycle, i.e. convert helium to carbon and oxygen, while the stars closer to eight Solar masses will be able to go through the third cycle, i.e. convert carbon to neon and magnesium. After that the nuclear energy generation will stop and the stars will collapse under gravity.

The large amount of energy released during each contraction of the central region of the star leads expansion of the outer gaseous layers of the star, which again contract when a new state of equilibrium is reached. In the expanded state, the outer layer is relatively cool and appears to be reddish in colour. In this expanded state, the radius of the star is very large and the star is known as a *red giant*. When the last cycle of energy generation is completed, the expanded outer layers are thrown out in a giant explosion. In this phase, the star is known as a *planetary nebula*. The name was given as early astronomers thought the large spheres of gas to be planets. A typical planetary nebula at a distance of 700 light years from us is shown in Fig. 5.1. The

Fig. 5.1 This planetary nebula, named the helix nebula, is situated at a distance of about 700 light years from the Earth. The central star is still in the process of becoming a white dwarf. Image credit: ESO

explosion is triggered by the energy released in the rapid contraction of the core, which collapses to a white dwarf, which we will describe below.

5.4.1.1 White Dwarfs

The stellar core is very hot and therefore all the atoms in it are completely ionised, that is, the electrons are all separated from the atomic nuclei around which they normally orbit to form atoms. During the collapse of the stellar core, the density keeps increasing and a stage comes when the free electrons come so close to each other that quantum mechanics, which is a branch of physics applicable to molecules, atoms and smaller entities, comes into picture. As noted in Sect. 5.2.1, for ordinary gas the pressure is directly proportional to the temperature. But at the high densities reached in the collapsing stars, quantum effects dominate and the electrons are said to become *degenerate*. In this case, the electrons on the average have much higher velocities than would be indicated by their temperature. The electrons then do not have properties of ordinary gas and one result is that the pressure that they exert is no longer proportional to the temperature. For a *degenerate gas*, it is possible to say approximately that the pressure depends only on the density. It is then known as the *degeneracy pressure*. As the collapse continues, a stage is reached when this pressure balances gravity and hydrostatic equilibrium is again reached. Since the nuclear fuel in the object is exhausted, it no longer a star, in the sense that it does not generate any energy through nuclear processes. But it continues to radiate its stored heat energy, as a result of which it cools over a long period of time. But the consequent decrease in temperature does not lead to a decrease in pressure, as the pressure now depends only on the density, because of the degeneracy. The balance between pressure and gravity therefore in effect lasts for ever.

When the object reaches a degeneracy pressure supported state, it is very dense and small in size. At birth such an object is very hot and appears blue-white, like very massive and hot stars (see Sect. 2.3.3) and is called a white dwarf. As the white dwarf cools, the colour progressively changes to red and eventually, in a few billion years, it becomes an invisible black dwarf. But there is no change in their size, mass and chemical content. This stage is therefore the end stage in the life of all stars with masses smaller than about eight Solar masses.

The typical mass of a white dwarf is comparable to the mass of the Sun, while its size can be several hundred times smaller. The density is therefore so high that just a spoonful of white dwarf matter can amount to a mass of a few tonnes (1 tonne = 1000 kg), which is comparable to the mass of an Asian elephant. How much mass of the original star is left in these collapsed cores or white dwarfs? All the observed white dwarfs have been found to have masses smaller than about 1.4 Solar masses. Why is this so? When the original stars have masses up to about eight Solar masses, why do all white dwarfs have these small masses, the rest of the mass being thrown out in the planetary nebular stage?

The answer to this question was first obtained by the Indian astrophysicist, Subrahmanyan Chandrasekhar in the early 1930s. Using the then recently established quantum mechanics and special theory of relativity, he proved that there is a limit to how much mass can be supported by the degenerate electron pressure. The reason for the existence of the limit is that as the mass of the white dwarf increases, its radius decreases and its density increases. The degeneracy pressure increases too, but because of the effects of special relativity, it is no longer sufficient to balance gravity. As proved by Chandrasekhar, this stage is reached when the mass of the white dwarf is 1.44 times the mass of the Sun. This is called the *Chandrasekhar limit* and all white dwarfs are required to have a mass smaller than this limit. We will see below what happens to a collapsing object which has greater mass.

As we have mentioned above, before reaching the white dwarf stage, the evolving stars go through a red giant phase when, as the name suggests, they will be some what cooler, and hence redder than their original colour. They will also be very large in size, which could reach to 200 times the radius of the Sun, which is the reason they are called giants. As one can see, this is the fate that awaits our Sun, of course after another 4.5 billion years. During the red giant stage, the Sun may be big enough to gobble up Mercury and Venus and possibly the Earth and Mars. On becoming a white dwarf, its size will be about the same as the size of the Earth. Its mass will remain close to what it is now and hence its density will be huge, as mentioned above.

Chandrasekhar

Subrahmanyan Chandrasekhar (1910–1995) was born in Lahore in a cultured Tamil family. At the time of his birth, his father was Deputy Auditor General of the Northwestern Railways. His mother was well educated. She had translated Ibsen's famous play 'A doll's house' into Tamil. It is said that his mother was responsible for arousing Chandrasekhar's curiosity at an early age. Nobel prize winner Sir C. V. Raman was Chandrasekhar's uncle. Chandrasekhar's family settled in Chennai when he was 8 years old. Chandrasekhar was home schooled till the age of 12. Later he did his graduation from the Presidency College, Chennai. He wrote his first research paper during that time. In 1930, he was awarded a government fellowship to pursue doctoral studies at Cambridge. On the way to Cambridge on a ship he calculated the upper limit on the mass of white dwarf stars. He improved upon it on reaching England, where he obtained his doctorate in 1933.

At that time, the famous astronomer, Sir Arthur Eddington was Plumian Professor at Cambridge and was regarded as one of the world's foremost astrophysicists. He publicly ridiculed Chandrasekhar's idea. In frustration, Chandrasekhar left England for the USA. Initially, he worked in the Yerkes Observatory. Later he shifted to the University of Chicago where he remained active till his death. In time, his work on white dwarf stars was accepted by all and he was awarded the Nobel Prize in Physics in 1983.

Chandrasekhar worked in several areas of astrophysics. He used to study a topic in depth, publish research papers on it and finally write a masterly book on the subject before moving on to a new topic of study. He has authored ten scholarly books and several lecture notes. He rewrote Newton's Principia in a way which will be understandable to a general reader. He was also fond of literature and arts along with science and has written a book titled Truth and Beauty: Aesthetics and Motivations in Science.

5.4.2 Stars with Mass Greater Than Eight and Smaller Than Twenty Five Solar Masses

For the generation of gravitational waves, we have to depend on massive objects of very compact size, so that they have a strong gravitational field. White dwarfs, even though they have unusually high density, are not quite sufficient for the purpose, and we have to go to objects with vastly greater density than white dwarfs, which are the neutron stars and black holes. These objects are formed at the end point of evolution of stars with mass greater than about eight Solar masses.

For stars with mass greater than eight Solar masses, the nuclear cycles mentioned in Sect. 5.2.2 continue all the way, through the fusion of hydrogen, helium, carbon, oxygen, magnesium, silicon to iron, releasing energy at every step. But it is not possible to go to nuclei heavier than iron in the same way. Fusion of iron requires energy to be supplied for the process, rather energy being released. As a result, once the central region of a star is converted to iron, no further energy releasing fusion is possible, however high the temperature and density may be.

As the energy generation stops on the formation of iron, the star starts cooling and thus contracting. The density and also temperature start increasing as a result of this contraction. The important point here is that for stars with an initial mass greater than eight Solar masses, which we are considering here, the mass of the collapsing core is greater than the Chandrasekhar limit of about 1.4 times the mass of the Sun. A white dwarf cannot therefore form, and the collapse continues beyond the point at which white dwarf densities are reached. When the collapsing object reaches a radius of a few tens of km, the nuclei of iron and other elements which are present cannot sustain themselves under the tremendous density and pressure reached in the core of the star. The nuclei therefore disintegrate into their constituent elementary particles, i.e. the neutrons, protons and electrons. The protons and electrons are very close and combine together to form free neutrons through an elementary particle process known as inverse beta decay. At this stage, the matter consists of free neutrons and a small fraction of protons and electrons. All the particles are degenerate due to the very high density of the matter. At normal densities, any free neutron would decay in about 12 min to a proton, electron and anti-neutrino. But this process, called beta decay, is suppressed in the very dense core, as the electrons are degenerate. In a broad sense, there is nowhere left for the electrons produced to go, so the neutron is prevented from decaying.

Extremely large amounts of energy, which can be greater than several times the energy given out by an entire galaxy in a second, are generated during the collapse of the core of the star. Because of this, the outer parts of the star are thrown out in a gigantic explosion in a spectacular fashion. Such a stellar explosion is called a *core-collapse supernova* and the compact remnant left behind is called a neutron star, which we will describe in the next section.

We can get an idea of the dynamics involved in such explosions from the following. In 1054 AD, it was noted in Chinese chronicles that a 'guest star' appeared in the sky. The star could be seen during daytime too, for several days before fading out. While the Chinese could see the object with their naked eyes, we now need a telescope to observe the object at the same position in the sky, which has been named the Crab nebula because of its appearance. This is shown in Fig. 5.2. What we see there are gases which are still dispersing as a result of the stellar explosion that was seen by the Chinese astronomers. These gaseous filaments are moving with velocities of about a few thousand km/s even after more than a 1000 years of the stellar explosion. We can recall that the fastest speed we, at times, drive on highways is not too much larger than 100 km/h! Thus, one can get some idea of the velocities and accelerations that these gases must have had at the time of the explosion. The structure that we see now is known as a *supernova remnant*, which constitutes the remains of a supernova explosion. While the supernova was observed on the Earth in 1054 AD, it actually happened 6523 years before that date, which we know since the distance to the site

Fig. 5.2 Crab nebula. This is the remnant of the supernova explosion which was witnessed by Chinese astronomers in 1054 AD. What we see are the gases still expanding at velocities exceeding 1000 km/s as a result of the explosion. Image credit: ESO

of the explosion is known to be 6523 light years from us. Theoretical studies suggest that the progenitor star which exploded as a supernova had a mass in the range 8–11 Solar masses.

5.4.2.1 Neutron Stars

We have seen above that the evolution of stars with mass greater than eight Solar masses leads to the formation of a core which is made up almost exclusively of neutrons. With further contraction of the core, the density of neutron gas becomes very high and this gas becomes degenerate similar to the electron gas in white dwarfs. The pressure of the neutron gas does not depend on the temperature any longer and once it balances the gravitational force, the star achieves equilibrium, in spite of the fact that it keeps losing thermal energy to the surroundings. So the equilibrium is effectively maintained forever, even when the temperature of the star tends to absolute zero. The collapsed core which is now in hydrostatic equilibrium is known as a *neutron star*. The formation of a neutron star is the end state for all stars having initial mass in the range of about 8 to 25 Solar masses.

The physics that applied to white dwarfs applies to neutron stars as well, and there is a limit for neutron stars, like the Chandrasekhar limit for white dwarfs, on the maximum mass that they can have. The value of this limit, called the *Tolman-Oppenheimer-Volkoff limit* (in short the TOV limit) after the physicists who discovered it, cannot be calculated exactly as there is still some uncertainty in our understanding of the behaviour of neutron star matter. However, the TOV limit is typically between two and three times the mass of the Sun.

Neutron stars have sizes of about 10–20 km and a mass quite often close to the Chandrasekhar limit of 1.4 times the mass of the Sun. Naturally, the density is extremely high, about a million, million kg per cubic centimetre. This is so high that just spoonful of neutron star matter on Earth would weigh as much as all of humanity. The neutron stars are dead stars in the sense that no energy generation takes place inside them. Like white dwarfs, even though they lose whatever energy they have at birth, by way of their being extremely hot, to the cooler surroundings, their size, mass and composition remain the same, as even cold neutrons can balance the attractive force of gravity due to their degeneracy pressure.

Two important properties of a neutron star are (i) it has an extremely strong magnetic field at the time of formation, up to 10^{16} times that of the Earth and up to about a billion times stronger than any that has been produced in laboratories on the Earth and (ii) at formation it spins very rapidly around itself and therefore has a very small rotation period. As we know the Earth spins once around itself in every 23 hours and 56 min, while the Sun has a rotation period of about 27 days. In contrast, at birth, neutron star periods are in the range of milliseconds (ms) to tens of milliseconds, but can later slow down to periods of some seconds. Both the strong magnetic field and rapid spin have to do with the fact that the star has shrunk by a factor of about a few hundred thousand from its original size and so the magnetic field and the spin rate have been magnified by factors of 10^{12} and a few million times, respectively.

The magnetic field of a neutron star is like that of a bar magnet and so the star has a magnetic axis, i.e. line joining the north and south poles of the magnet. The spin axis, that is, the axis around which the neutron star spins is often not aligned with the magnetic axis. The rotation of the neutron star therefore leads to an off-axis rotation of the magnetic field associated with it, which in turn leads to the emission of electromagnetic radiation. The emitted radiations has a specific broad pattern known as dipole radiation. But what is very interesting is that a small fraction of the emitted energy is channelled into two narrow cone-shaped beams aligned along the magnetic axis as shown in Fig. 5.3. As the neutron star rotates, the beams sweep the sky along circles similar to the search light of a circus party. If the Earth happens to be somewhere along the circle, it will receive a pulse of radiation once every rotation as the beam passes through the site line of an observer on the Earth. Over a length of time the observer will see a series of pulses with the time between successive pulses being equal to the rotation period of the star. A neutron star from which such pulses are received is known as a *pulsar*. Some of the radiation is emitted at radio wavelengths, where the pulses are easiest to observe using a radio telescope. Pulses are also seen in the optical, X-ray and other regions of the spectrum.

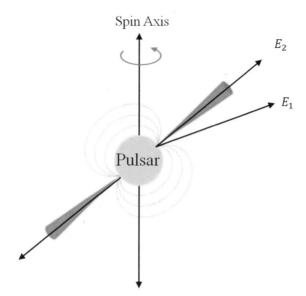

Fig. 5.3 A rotating neutron star with a strong magnetic field, which is observed as a pulsar. The axis of the magnetic field is along E_2. This is not aligned with the spin axis. The blue lines are the magnetic lines of force. As explained in the text, a fraction of the energy emitted by the pulsar is directed into two narrow cones as shown, at radio and other wavelengths. E_1 indicates the direction of the observer. Because of the rotation of the pulsar, the beams sweep the sky. If a beam happens to sweep through the line of sight of a radio telescope on the Earth, then a pulse of radio radiation is observed from the pulsar each time the beam sweeps through. Image credit: Kaushal Sharma

White Dwarfs to Neutron Stars

When a star has exhausted all its sources of nuclear energy, the interior cools, its pressure decreases and it can no longer sustain the inward force of gravity. The interior region then collapses forming a white dwarf, neutron star or black hole depending on its mass. A white dwarf is formed when the mass of the collapsing region is less than the Chandrasekhar limit. After formation, a white dwarf gradually cools by emitting radiation and over a very long period of time disappears from view. But the fate of a white dwarf is quite different if it happens to be in a binary system, with the companion being a normal star. In such a case, matter can be transferred from the companion star to the white dwarf and increase its mass beyond the Chandrasekhar limit. In this case, the white dwarf can collapse to form a very rapidly spinning neutron star with a high magnetic field, which can be a millisecond pulsar (a pulsar with rotation period of a few milliseconds). But in most cases, it is expected that the transfer of matter to the white dwarf increases the temperature in its interiors very much, triggering very rapid nuclear burning. This leads to an explosion which is known as Supernova

type Ia, with no compact object left behind. These supernovae have properties
which are very useful in cosmological investigations.

A radio pulsar was first discovered by Anthony Hewish and his research student
Jocelyn Bell in 1967. Soon after the discovery it was established that the object
emitting radio pulses must be a neutron star. A very useful characteristic is that
the pulse period is very nearly constant because the neutron star is so massive and
rotates so rapidly that it is very difficult to perturb the rotation rate. As the pulsar emits
radiation, it gradually loses energy and therefore the rotation slows down, leading to
a very slow increase in the rotation period. This is so slow that the period increases by
only a ten millionth of a second or even less in a year. Pulsars are, therefore, among
the most accurate clocks available for the measurement of time. We will see later
that the accurate pulse periods can be used to establish the existence of gravitational
waves.

5.4.3   Stars with Mass Greater Than Twenty Five Solar Masses

If the original mass of a star is larger than about 25 times the mass of the Sun, it will
evolve in a way similar to the stars with masses roughly between 8 and 25 times the
mass of the Sun as described above and will undergo a supernova explosion. The
mass of the remnant made up of neutrons, left after the explosion, will be larger than
the TOV limit. It then cannot achieve equilibrium as the pressure of the degenerate
neutron gas cannot balance the huge gravitational force. The remnant is then doomed
to collapse forever. This is the fate of all stars with masses larger than about 25 times
the mass of the Sun. Once the size of the star becomes smaller than a certain limit
called the *Schwarzschild radius*, it becomes what we call a black hole which we
describe below.

5.4.3.1   Black Holes

As we all know, a ball thrown up from the surface of the Earth travels up to a certain
height, its velocity decreasing with time because of the gravitational pull of the
Earth. When the velocity becomes zero, the ball stops and then falls back on the
Earth because of the same pull. If we throw the ball with larger velocity, it will be
able to reach greater height before falling back. What if we throw the ball with very
high velocity so that the pull of the Earth will not be sufficient to stop it? Yes, this
is possible and in that case, the ball will continue to rise up, albeit with decreasing
velocity, but will not fall back on the Earth. The minimum velocity with which we
have to throw the ball for this to happen is 11.2 km/s. This value is called the *escape*

velocity from the Earth, meaning, it is the minimum velocity with which an object has to be thrown vertically upwards from the surface for it to permanently escape the gravitational pull of the Earth.

The value of 11.2 km/s is obtained by using the mass and the radius of the Earth. Every astronomical object has an escape velocity. For example, its value on the surface of the Moon is 2.4 km/s while that on the surface of the Sun is 618 km/s. Obviously, higher the mass, higher is the value because of higher gravitational pull. However, the more important factor for us here is the size of the object. For two objects with the same mass, the smaller the size, the higher is the value of escape velocity. Thus, as a star contracts, its escape velocity increases.

As we saw above, the remnants of supernova explosions of stars with masses larger than about 25 times the mass of the Sun cannot achieve a balance between pressure and gravitational force and are doomed to contract forever. In this process of contraction and consequently that of increasing escape velocity, a stage comes when the escape velocity becomes larger than the velocity of light in vacuum, i.e. 300,000 km/s. This means that no object with velocity smaller than or equal to that of light can escape the gravitational pull of the stellar remnant.

As we know from the special theory of relativity described in Sect. 2.4, nothing can travel faster than light, and therefore neither light nor information conveyed in any way can come out of this object. We can only see an object because of the light it emits or reflects. Thus, for all practical purposes, the star will become invisible and all we can see at its position will be a black hole, which justifies the name given to these stellar remnants.

Objects from which light may not be able to escape were first considered by John Michell and Pierre Simon Laplace more than two centuries ago, using Newton's theory of gravitation. The fact that light could not escape was then not a serious concern, because velocities greater than light were possible in Newton's theory and if not light other objects moving with higher velocities could come out. This situation changed with the advent of special relativity, because no signal could travel faster than light, so if light could not escape from an object, then it would not be possible to receive any signal at all from the object by the outside world. But we should recall here that special relativity and Newton's theory of gravity are not compatible and we need to use general relativity to examine situations in which gravity is involved.

In general relativity, the concept of a black hole emerged from the very first exact solution of Einstein's equation published by Karl Schwarzschild in 1916. This described the spherically symmetric gravitational field around a point object and it was found that there is a radius, later named the Schwarzschild radius, from within which light cannot escape to the outside world. The radius is proportional to the mass of the black hole. For a star of mass equal to the mass of the Sun, this radius is about 3 km. It can be imagined that the black hole is surrounded by a spherical surface with radius equal to the Schwarzschild radius. No signal can escape to the outside world from within this surface, and anything that enters this surface is forever trapped and inexorably falls to the centre. The surface is said to be an *event horizon*. The existence of the event horizon is an exact mathematical result of general relativity and is a consequence of the enormous curvature of space-time. In general relativity, a black hole represents a physical and mathematical singularity, where matter density and space-time curvature become infinitely large.

A black hole can also have rotation or spin, and in such a case it is known as a *Kerr black hole*, after Roy Kerr who discovered in 1963 the corresponding solution of Einstein's equations. For such a black hole, the shape of the event horizon is more complicated than the simple spherical shape obtained for the Schwarzschild solution. The shape depends on the mass as well as the spin of the black hole. It can be shown that a black hole can have only three properties, mass, spin and electric charge. We do not expect to find charged black holes in nature, as stars are electrically neutral. When the spin is zero, we have the Schwarzschild black hole, and when the spin is non-zero we have the Kerr black hole. So a black hole is a pristinely simple object. Nevertheless, black holes are physically and mathematically extraordinarily interesting, and are the key to much of modern astrophysics.

Even though black holes cannot be seen directly, plenty of indirect evidence has been obtained for their existence. One important evidence comes from cosmic sources which are strong X-ray emitters. Astronomers have been able to understand the nature of these sources from their detailed study. Some of these X-ray sources have been identified with binary systems consisting of two stars which are gravitationally bound to each other, and go around their common centre of gravity as seen in Chap. 3. These are called X-ray binaries. One member of these systems is an ordinary star, while the other is a compact object like a neutron star or a black hole. X-rays are emitted when material from the ordinary star gets attracted towards the compact object because of its strong gravitational pull, gets extremely heated in the process and emits X-rays (see Sect. 2.3.3). Analysis indicates that for some of these X-ray sources, the mass of the compact object is higher than the TOV limit for neutron stars and hence it cannot be a neutron star. It must then be a black hole. Observations are consistent with its being a black hole. Such black holes have been observed to have masses between 5 and 15 Solar masses.

White Holes

A black hole is an object with non-zero mass but zero size. We have seen that a black hole can be imagined to be surrounded by a surface known as the event horizon. Any body which enters the event horizon can never escape to the outside and is lost for ever. This is also true with light, which gives the black hole its name. A white hole can be viewed as the reverse of a black hole. It too has a mass concentrated at a single point and is surrounded by a horizon. In this case, matter can escape from inside the event horizon to the outside, but cannot enter the horizon. So, a white hole can be a source of energetic particles and radiation. The idea of white holes first arose in the mathematical investigation of the geometry of space and time around black holes. While white holes have interesting properties, it is not known how they could be formed. While black holes can naturally form in the collapse of stars, there is no such natural process

known which could lead to the formation of white holes. It is speculated that
if at all they exist, they should have been around since the big bang.

Evidence has also been obtained for black holes which have masses exceeding
about a million to hundreds of million times the Solar mass. These supermassive
black holes lie in the centres of most galaxies including our own Galaxy and quasars.
The evidence for these being black holes comes from the huge amount of energy
emitted by quasars from a very compact region which can only be explained as energy
emitted by matter falling on to a supermassive black hole. Such a black hole with a
mass of 4.3 million times the mass of the Sun is believed to be present at the centre
of the milky way.

Galaxies
A faint, whitish band is seen stretching from horizon to horizon in the night
sky when observed on a dark night away from city lights. The band is known
as the Milky Way. Galileo Galilei, around the year 1609, first observed the
faint band with his small telescope and found that it is made up of a large
number of faint stars. It was later realised that the band represents the distant
parts of a structure which is called the Milky Way galaxy, in which the Sun
is located. It was established in the first few decades of the twentieth century
that the Universe contains a large number of galaxies, which, in fact, are the
basic building blocks of the Universe. A galaxy like the Milky Way contains
about a hundred billion stars, but there are galaxies which are smaller and
contain far fewer stars, and others which can contain a hundred times as many
stars. The stars in the Milky Way are distributed mainly in a disc-like structure,
which has a diameter of about 100,000 light years. Towards the central region
there is a bulge, and within the body of the disc there are conspicuous spiral
structures, known as spiral arms. About half of the galaxies have shapes similar
to the Milky Way galaxy and are known as spiral galaxies. Another major type,
elliptical galaxies, are quite different in shape. When observed with a telescope,
they appear to have a shape like an ellipse and are known to be ellipsoids in
three dimensions. These galaxies mostly have older stars, while the discs of
spiral galaxies have younger stars in them, formed from the cold gas and dust
that are present in the disc.

Cosmological Redshift
It was first observed by the astronomer Vesto Slipher in 1912 that light from
distant galaxies is redshifted. This means that the wavelength of absorption
lines in the spectrum of light from the galaxies is increased relative to the

wavelength of the corresponding lines in the laboratory. The redshift can be interpreted as being due to the Doppler effect described in Sect. 2.5.6, from which it follows that the redshifted galaxies must be receding from us with a high velocity. In the 1920s, it was discovered independently by Georges Lemaître and Edwin Hubble that the recession velocities of the galaxies are proportional to their distances from us, which can be interpreted to mean that the whole Universe is expanding. This is known as the Hubble-Lemaître law and the redshift caused by this expansion is known as *cosmological redshift*. The relation between velocity and distance leads to the determination of a quantity known as *Hubble's constant* which is of great importance to cosmology.

When the measured redshift of a galaxy is significantly smaller than 1.0, it can be shown that that the redshift is approximately proportional to the galaxy's velocity of recession. It then follows from the Hubble-Lemaître law that the distance to a galaxy is proportional to its redshift. So if the redshift of a galaxy is known, its distance is determined. As the distance to the galaxies increases, this simple interpretation can no longer remain tenable. If it did, beyond some distance the velocity of recession would exceed the velocity of light, which would violate special relativity. But here general relativity comes to the rescue: as the distance increases, the curvature of space-time has to be taken into account and the simple relations used by Hubble and Lemaître no longer apply. More complicated relations derived from general relativity have to be taken into account. Using these relations, given the redshift of a galaxy its distance can be determined. But the redshift is no longer proportional to the velocity and conflict with special relativity is avoided.

5.5 Sources of Gravitational Waves

Here we will briefly discuss some sources of gravitational waves which result from the evolution of stars. These sources are mainly of four types: (1) burst sources, like supernova explosions and *Gamma-ray bursts*, which are short duration events; (2) continuous sources, like spinning, compact objects which can emit gravitational waves at the same frequency for a long duration; (3) binary sources with the two components being the compact objects white dwarfs, neutron stars or black holes and (4) stochastic background sources which are collections of a large number of weak sources, which cannot individually be detected but can be collectively observed as a background of gravitational waves.

Supernovae and Gamma-ray Bursts The collapse of a stationary interstellar cloud results in a perfectly spherical star. However, most clouds are rotating and that causes the shape of the star to depart from being a perfect sphere. This is similar to the Earth being flatter at the poles compared to the equator. This, in addition to other reasons, like the presence of magnetic fields, convection, etc, can result in a core-collapse supernova explosion (see Sect. 5.4.2) which is not spherically symmetric. Observations of supernova remnants, like the Crab nebula, and the position and velocity of the remnant neutron stars/pulsars point to the fact that the explosions are not spherically symmetric. The material of the star undergoes very high acceleration at the time of explosion, as seen from the velocities of the material of the Crab nebula as well as from theoretical understanding.

Only about 10^{-10} of the mass lost in the process is emitted as gravitational energy, but this energy is still large enough to make such an event detectable by *Advanced LIGO* to a distance of about 30,000 light years. The emission of gravitational waves from supernovae will last only from a few ms to a few seconds. The frequency of the gravitational waves is in the range of a few to about 10^3 Hz, which is in the range detectable by Advanced LIGO. The probability of observing a core-collapse supernova happening in our galaxy within a distance of about 30,000 light years is estimated to be a few times 0.1 per decade, so such a explosion could eventually be observed. The deployment of more sensitive gravitational wave detectors in the future, as described in Chap. 9, will make it easier to observe such supernova explosions.

Gamma-ray bursts are events in which a tremendous amount of energy is released. In such a burst, first there is explosive release of Gamma rays, which is followed by electromagnetic radiation of various kinds, including X-rays, optical radiation and radio waves. Thousands of Gamma-ray bursts lasting for various durations, ranging from a fraction of a second to hundreds of seconds and longer, have so far been observed. The amount of energy emitted in such a short time is equivalent to the total energy emitted by a star like the Sun in about a trillion years. The bursts are mainly of two types: (1) long duration bursts which last for longer than about 2 s, whose origin is believed to be in core-collapse supernovae of a rapidly rotating star. Such an event will emit gravitational waves of the type described above and (2) short duration bursts of less than about 2 s, which are believed to originate in the merger of two neutron stars in a binary system. Gravitational wave detection from a merger of this type quickly followed by the detection of the associated Gamma-ray burst is described in Sect. 8.5.

Pulsars: The spin period of pulsars increases with time, i.e. the pulsars slow down as a result of their energy emission. This provides the acceleration (rather deceleration) necessary for generation of gravitational waves. Thus, a pulsar can be a potential source of gravitational waves if it has some asymmetry because of some deformities, like a large mountain on the surface of the pulsar. Such pulsars will emit gravitational waves continuously at nearly the same frequency, typically in the range of a few Hz to about 10^3 Hz. Since the signal is continuous with the same frequency, it is possible to add the signal obtained over a period of time, thus making it easier to detect weak signals.

Binary Neutron Stars: About half of the stars that we see in the sky are members of binary systems. Each such system consists of two stars which are gravitationally bound to each other and are orbiting around each other. We have already come across binary systems in Sect. 2.5. Binary sources are like rotating dumbbells (see Fig. 4.2) and can emit gravitational waves. The faster the rotation, the stronger will be the gravitational wave emission.

A binary system with two neutron stars can be very compact, since each neutron star is a very compact object, with radius typically about 10 km. Such a binary will have very high rotation speed and will be a strong source of gravitational waves. As the system loses energy through emission of gravitational waves, the orbit will shrink and the frequency and amplitude of the waves will increase with time. In the initial stages of this *spiral-in*, the distance between the two neutron stars is large. The orbital period, i.e. the time taken by the neutron stars to go once around each other is therefore large, and so the frequency of the *orbital motion*, i.e. the number of times the two objects go around each other per second is small. The frequency of the gravitational waves emitted is twice that of the orbital frequency (see Sect. 8.6 for exceptions to this rule). As the neutron stars spiral-in, the orbital frequency increases, and the maximum orbital frequency, and therefore the maximum gravitational wave frequency is obtained when the neutrons stars just touch each other. This maximum frequency is in the range of 1000–2000 Hz. When the neutrons stars get very close they tear each other apart due to the great gravitational force, merge together and become one entity. The nature of the entity formed from the merger is decided by its mass. If the mass is smaller than the TOV mass limit, it will form a neutron star, otherwise it will become a black hole. During the merger, strong gravitational waves are emitted.

Indirect observational evidence for the emission of gravitational waves from an in-spiralling binary neutron star system was obtained in the early 1970s and is described in Chap. 6. The emission of gravitational waves from such a system and the merger of the neutron stars have also been detected and will be described in Sect. 8.5.

Binary Black Holes: Black hole binaries, i.e. two black holes orbiting each other will be strong emitters of gravitational waves. They will spiral-in because they will be losing energy due to the emission and will eventually merge forming a single black hole, producing a strong characteristic signal of gravitational waves which will last for several ms. In such a system, the maximum gravitational wave frequency may be imagined to occur when the event horizons of the black holes touch each other, but in practice it can occur before that due to certain features in the orbit. The greater the mass of the black holes, the smaller will be the maximum frequency. The maximum frequency also depends on whether the black holes are themselves rotating and how this rotation is aligned with the orbital motion. The maximum frequency is expected to be in the range of a few thousand Hz. The first direct detection of gravitational waves was of such an in-spiralling and merging black hole binary. This event, and other events like it detected by Advanced LIGO and *Advanced VIRGO* will be described in Chap. 8.

Neutron Star-Black Hole Binary: It is possible that a binary system has a black hole and a neutron star as its members. Such a binary will also be a strong source of gravitational waves. There have been candidate events which could be such systems, as described in Chap. 8. Confirmation is awaited at the time of writing.

Stochastic Background Sources: There can be a very large number of sources of gravitational waves in our galaxy, in other galaxies and in the early Universe even before galaxies were formed. Each of these sources can individually be too faint to be detected, either because it is is not a strong emitter of gravitational waves or because it is at very great distance. Examples of such sources are white dwarf binaries. The mass of white dwarfs is similar to the mass of neutron stars, but the white dwarfs are much larger in size, with a radius of several thousand km. Therefore, the size of white dwarf binaries is much larger than that of neutron star binaries, which results in much smaller energy being emitted as gravitational waves. But there can be hundreds of millions of such faint white dwarf binaries in our galaxy emitting at frequencies greater than about 10^{-4} Hz. The emission from such sources, and other similar sources, can add up to form a background of gravitational waves which is said to be stochastic, because it fluctuates randomly with time, but can be studied statistically. The gravitational wave frequency of this background is simply too low for observation by the detectors presently available on the Earth, but would be detectable in the future from space, as described in Chap. 9. The stochastic background can have contributions from processes taking place in the very early Universe. The gravitational waves produced in the very early Universe can also be detected in principle through the effect they have on the observed *polarisation* of the *cosmic microwave background radiation*, which is the presently observed remnant of the hot radiation produced in the big bang.

Chapter 6
Evidence for the Existence
of Gravitational Waves: The Binary
Pulsar

Abstract In this chapter, we describe how the first indirect evidence for the existence of gravitational waves came from a binary system having both components as neutron stars. The binary was discovered by Hulse and Taylor in 1974 through the detection of radio pulses emitted by one of the neutron stars. Detailed observations yielded information about the masses of the components and the rotation period of the binary system. It was soon realised that the system would lose energy through emission of gravitational waves. Loss of energy would cause shrinking of their orbits and consequently, a decrease in the period of rotation of the binary. Hulse and Taylor and later Taylor and Weisberg continued observing the binary for several years. Their observations indeed showed a decrease in the rotation period and led to the accurate determination of many physical parameters related to the system. The rate of decrease matched, to within one percent, theoretical predictions made using Einstein's general theory of relativity. Hulse and Taylor were awarded the Nobel Prize in physics for the year 1993 for their work.

6.1 Introduction

Even though gravitational waves were predicted a hundred years ago, they remained a mere theoretical possibility until the 1970s. We have seen earlier that no object on the Earth is capable of producing gravitational waves which are sufficiently strong to be detected by the gravitational wave detectors already built or to be built in the foreseeable future. The detector technology was not developed sufficiently until 2015 even for detecting gravitational waves generated by compact astronomical sources discussed in Chap. 5, unless the sources happened to be not too far in our galaxy. We will learn about the history and the present status of the detectors in Chap. 7. Here we will describe how the first real proof of the existence of these waves was obtained indirectly through the study of a binary pulsar.

A. Kembhavi and P. Khare, *Gravitational Waves*,
https://doi.org/10.1007/978-981-15-5709-5_6

6.2   Detection of Pulsars

A lone neutron star would be hard to detect, because it does not emit much radiation. But we can observe a lone neutron star when it is a radio pulsar, for which it needs to have a high magnetic field and rapid rotation, which results in radio pulses and gives the radio pulsar its name. We have seen in Sect. 5.4.2.1 that the pulses are emitted in a narrow beam which sweeps the sky as the neutron star rotates, and we will detect a pulsar only if we are located along the sweep of the beam. Since this happens only for a small fraction of all pulsars, the number of radio pulsars we see is only a fraction of the actual number of pulsars in the galaxy. As a pulsar ages, its rotation slows down because of loss of energy, and the magnetic field decays in strength, as a result of which a stage is reached when emission of pulses is no longer possible. There should be many such non-pulsating lone neutron stars in our galaxy which we are not able to detect.

A neutron star in a binary system can be detected when it is a radio pulsar or when the companion is a star from which matter is flowing onto the neutron star, in which case X-rays are emitted. In the latter case, radio pulses cannot be detected even if they are being emitted, because they are absorbed by the matter flowing onto the neutron star. In B1913+16, which is a pulsar in a binary system as we will see below, the companion of the pulsar is also believed to be a neutron star. No radio pulses are being detected from the companion, which could be either because we are not in the sweep of the beam, or the neutron star is old and is not emitting pulses. It is believed that the latter is the correct explanation for B1913+16. Some cases of double pulsars, where both the objects in a *compact binary* are radio pulsars, have been observed.

The number of binary neutron stars we detect is only a small fraction of the total number of such binaries in the galaxy, since we detect such a system only when we can observe the radio pulses from either one or both of the neutron stars. A system in which neither of the neutron stars is a pulsar is hard to detect by conventional means. Such a system emits gravitational waves, which can be detected in principle, but that had not been possible until recently. When one of the two neutron stars is a pulsar, very accurate observations are possible from which the emission of gravitational waves can be inferred. We will see below how this is done in the case of B1913+16.

6.3   Formation of the Binary Pulsar

In Sec. 5.4.2, we described how neutron stars are formed at the end of the evolution of massive stars whose mass is greater than about eight times the mass of the Sun. When such a star exhausts all the nuclear fuel available to it, the centre collapses to form a neutron star, while the outer parts are thrown out in a supernova explosion. So we can imagine that a binary with two neutron stars began as a system of two massive stars in orbit around each other. A neutron star would form when each star explodes at the end of its life, so that after two explosions we have a system of two neutron stars, one of which is a radio pulsar. But there are difficulties with this simple picture.

As discussed in Sect. 5.3, the lifetime of a star depends on its mass, and the greater the mass, the shorter is the lifetime. In a binary composed of two massive stars, say A and B, the star with the greater mass, say A, will evolve faster, and having reached the end of its life will explode as a supernova. Most of the mass of the star is lost in the explosion, and the neutron star that is left behind has much smaller mass, close to the Chandrasekhar limit of 1.4 times the mass of the Sun. The gravitational attraction between the remnant neutron star and star B is no longer enough for the system to remain as a binary held together by gravitational attraction. The binary is therefore disrupted and the two stars go on their own way. So how do we form the binary pulsar?

The answer lies in the exchange of mass between the two stars making up the binary system. Since star A is the more massive star, it evolves first, its interior region contracts and the outer region expands, as described in Sect. 5.4. If the binary is compact, i.e. the distance between the stars is sufficiently small, when star A has expanded enough, matter flows from star A to star B. This loss of mass from star A can be so much that star B becomes the more massive star. Star A continues to evolve until it explodes as a supernova, leaving behind a neutron star. But the explosion does not disrupt the binary, because now *the exploding star has lesser mass than the companion star*. The result, therefore, is a binary system consisting of a neutron star which can be a radio pulsar, and the ordinary star B.

In the binary, star B follows its own evolution and when it in turn expands, matter from it can flow to the neutron star. This releases large amounts of energy, some of which is emitted in the form of X-rays, thus giving rise to an *X-ray binary* system. As the evolution continues, in favourable circumstances a great deal of matter of star B is lost from the binary system, so that when star B explodes at the end of its evolution, another neutron star is formed without disrupting the system. At this stage, the binary consists of two neutrons stars, either of which or both can be radio pulsars. If the mass of star B remains greater than the mass of the companion neutron star, then the system is disrupted when star B explodes, with the two neutron stars speeding away in the galaxy. Many variants of these processes are possible, leading to a variety of binary systems of compact objects. The processes of mass exchange and mass loss are very complex and difficult to study in detail. Such mass exchange between companion stars in a binary was first proposed in 1955 by A. Crawford, who was a young research student in Britain, and independently by the great British astronomer Fred Hoyle.

6.4 Discovery of the Binary Pulsar

In 1974 Joseph Taylor of Princeton University and his graduate student Russell A. Hulse, carried out a very sensitive survey for observing new pulsars. It was then only seven years since the discovery of the first radio pulsar and only about 100 pulsars had been observed. Just to compare, more than 2700 radio pulsars have been observed until the end of 2019. Until 1974 it had not been possible to determine the mass of any of these pulsars. We have seen in Chap. 2 that it is possible to determine the masses of stars which are members of a binary. Hulse and Taylor were hoping that in their survey they would be able to find a pulsar which is a member of a binary and that they would be able to determine its mass. For their survey, they were using the largest radio telescope available at that time which was located at Arecibo in Puerto Rico. They also used computers to aid their search and their detection capability was ten times better than earlier observations. They discovered a pulsar on July 2, 1974. It had a rotation period of 59 ms and was the second-fastest rotating pulsar at that time, the fastest being the Crab nebula pulsar with a period of 39 ms. It was named PSR B1913+16 according to its position in the sky.

Continued observations and analysis revealed that the period of the pulsar was changing from day to day. This had not been observed for any of the 100 or so pulsars observed till then, all of which had rotation periods which increased extremely slowly with time, along with occasional unpredictable small changes. A careful study of the observed data (see Sect. 6.5) revealed that the apparent change in rotation period was occurring due to the fact that the pulsar was a member of a binary system with the two stars going around each other every eight hours. The other member of the binary is also a neutron star but we can not see it as a pulsar. An artistic impression of the binary pulsar is shown in Fig. 6.1.

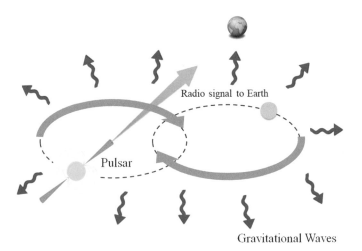

Fig. 6.1 Artistic impression of the binary pulsar discovered by Hulse and Taylor. One of the two stars is a pulsar. The other too is a neutron star but not a pulsar. The pulsar emits radio waves and other radiation in narrow cones as usual, while the binary emits gravitational waves in a broad pattern. Image credit: Kaushal Sharma

6.4.1 The Nature of the Binary Pulsar

Each of the two neutron stars in the binary above has a mass of about 1.4 times the mass of the Sun, and radius of about 10 km. Because of the second explosion, the binary has a very eccentric orbit, so each neutron star moves around the companion neutron star along a path which is a narrow ellipse, as shown in Fig. 6.1. The minimum distance between the two neutron stars, which is known as the *perihelion* distance, is very much smaller than the maximum distance between them. Such a small minimum distance is possible only because both neutron stars have a small radius, which permits them to approach each other very closely. Normal stars, because of their much larger radius, would not be able to get so close without colliding and possibly getting destroyed.

The compact nature of the neutron star binary means that the gravitational force between them is very strong, particularly at the perihelion distance, and the neutron stars move with very high velocity along their orbital paths. Newton's theory of gravity cannot be applied to such a system and we instead need to use Einstein's general theory of relativity. We have seen in Sect. 3.5 that apart from the three famous tests of general relativity, there are very few phenomena, like gravitational lensing, where the effects of Einstein's theory can be clearly seen. In all such cases, the effects are very weak, and what we need is a system where the gravitational field is so strong that Einstein's theory can be further tested. A neutron star binary is just such a system.

6.5 Further Studies of B1913+16

The early observations by Hulse and Taylor established that B1913+16 was a binary consisting of a rapidly rotating radio pulsar and a companion. The nature of the companion was not immediately known, but it was clear that it should be a compact object with a small radius. If it were an extended object like a star, then as the pulsar went around it, the pulsar would be eclipsed, but this was not observed. The compact object could be a white dwarf, neutron star or black hole. But observations and theoretical work over the years have established that the companion is a neutron star, rather than any of the other possibilities. No radio pulses have been observed from the companion. Therefore either the radio beam from the companion does not sweep past the Earth, or it is not a pulsar. The fact that both components of the binary are so compact is important in the analysis of the system. If the companion had been an extended object, the gravitational field of the pulsar would have distorted the shape of the companion, and it would have been difficult to interpret the observations. The neutron stars in the binary do not have such distortions, and being compact, can approach each other very closely. The system,, therefore, is an ideal laboratory for the study of gravitational effects.

Hulse and Taylor and later Taylor and J. Weisberg continued to study B1913+16 for many years with the Arecibo radio telescope. A pulse is detected every time the radio beam from the pulsar sweeps across the telescope. The time between successive pulses is equal to the rotation period of the pulsar, which is very close to 59 ms. The rotation period remains very nearly constant over long periods of time, because the neutron star is such a massive object and it is very difficult to affect its rotation. The rotation of the pulsar does slow down measurably over long periods of time, since the pulsar loses energy because of the radiation that it emits.

Even though the pulsar rotation period changes very slowly, the time between successive pulses received on the Earth can change due to various other reasons. As the pulsar moves along its orbit around the companion star, it is sometimes moving towards the Earth, and at other times it is moving away. During its motion towards the Earth, the observed time between two successive pulses decreases and becomes less than the rotation period of the pulsar. When the pulsar is moving away, the observed time between the pulses increases and becomes greater than the rotation period. This is due to the Doppler effect described in Sect. 2.5.6. Using sophisticated instruments and measurement techniques, Taylor and Weisberg were able to measure the arrival time of successive pulses extremely accurately. This in turn led to very accurate velocity measurements. Such measurements, made over many orbits are shown in Fig. 6.2. It is the existence of such variation that establishes the binary nature of the object.

From arrival time measurements it is possible to infer many details about the orbit, including the size and shape of the orbit, how closely the pulsar approaches its companion and so forth. Such calculations have been done over the years for several binary star systems having components as ordinary stars using Newton's theory of gravitation. What is generally not possible in such calculations is the measurement of the masses of the stars in the binary. The masses can be estimated only in special circumstances and accurate values can seldom be obtained. The situation is quite

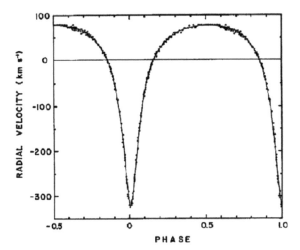

Fig. 6.2 The change in the velocity of the pulsar as it moves along an orbit. The x-axis shows the position along an orbit and the y-axis the velocity of the pulsar in the direction of the Earth. When the velocity is positive the pulsar is moving away from the Earth while negative velocity indicates that the pulsar is approaching the Earth. Image credit: R. A. Hulse and J. H. Taylor, Astrophys. Journ. **195**, L51, 1975

different for B1913+16. Here the gravitational field is very strong, and general relativistic effects, including gravitational redshift, and the precession of the perihelion, which we discussed in Sect. 3.5 are present. These lead to further changes in the pulse arrival times. When calculations are performed to account for all the effects, using general relativity, it turns out that all the parameters which are required to determine the shape of the orbit, as well as the mass of the pulsar and its companion can be very accurately determined.

The values for some important quantities calculated from the analysis of pulse arrival times are shown in Table 6.1. These are based on measurements made for more than thirty years after the discovery of the pulsar. The many places after the decimal point show how accurate the measurements are. The precession of the perihelion in the system is about 4.2 degrees per year. This is about 35,000 times the value for the precession of Mercury, which is only 43 arcseconds per century! This shows how strong the gravitational field is in the present case. The mass of the pulsar and the companion neutron star are 1.44 and 1.39 times the mass of the Sun, respectively. When such mass measurements were first reported some years after the discovery of the binary pulsar, they represented the first ever direct measurements of neutron star mass. The measurements are also the most accurately measured stellar mass ever made. The fact that the two masses are so close to the Chandrasekhar limit of about 1.4 times the Solar mass has implications for the way in which the neutron stars were formed.

The maximum distance between the pulsar and its companion is about 3.15 million km, which is very much smaller than the Earth-Sun distance of 150 million km. The minimum or perihelion distance between the two is 746,600 km, which is comparable to the radius of the Sun, which is 700,000 km. The two components of the binary can approach each other so closely only because they are very compact objects. If the companion were more extended, like a normal star, it would be disrupted if the distance of approach was so small. At the perihelion, the pulsar moves with a velocity of about 400 km/s. The time taken to complete one orbit, which is known as the orbital period, is about 7 h and 45 min. The binary system is at a distance of 21,000 light years away from us. Because of its position in the sky, the system will be

Table 6.1 Some important parameter values for the binary pulsar

Parameter	Value
Pulsar rotation period	59.0300032180 ms
Orbital period	7.7519938773864 h
Maximum separation	3,153,600 km
Minimum separation	746,600 km
Precsssion of perihelion	4.226598 degrees/year
Pulsar mass	1.4398 Solar Mass
Companion mass	1.3886 Solar Mass

visible only between 1941 and 2025. It was indeed fortunate that Hulse and Taylor could detect the system during this limited interval of time that is available for its observation from the Earth.

6.6 Proof for the Existence of Gravitational Waves

Soon after the discovery of the binary pulsar, it was pointed out by Robert Wagoner that the observation of change in the binary rotation period with time will allow us to test the presence of gravitational waves. The binary pulsar is very compact and the gravitational force between the two neutron stars is very strong. The orbit of the pulsar around the companion is highly elliptical in nature. As seen in Sect. 5.5 such a binary emits a significant amount of gravitational waves due to the high acceleration. The emission results in the loss of energy by the binary, and consequently the binary

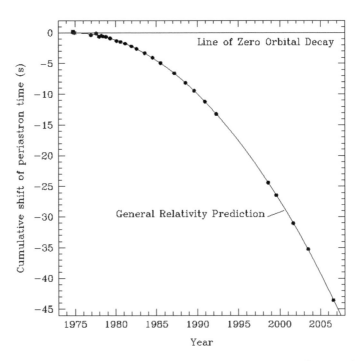

Fig. 6.3 The change in rotation period of the binary B1913+16 over a span of 30 years from 1975 to 2005 as predicted by Einstein's General theory of relativity and as observed. The x-axis shows the time of observation and the y-axis shows the decrease in time taken for successive revolutions, which indicates shrinking of the orbit. The dots show the observed values. The agreement between observed and predicted values is remarkable and proves the existence of gravitational waves. If gravitational waves were not being emitted, the graph would have been a straight line as shown at the top. Image credit: J. M. Weiberg, D. J. Nice and J. H. Taylor, Asrophys. Journ. **722**, 1030, 2010

shrinks in size. The two neutron stars come closer together and therefore revolve faster around each other, leading to a decrease in the period of the binary. A measurement of the change in their period of revolution allows a measurement of the rate at which they are losing energy through the emission of gravitational waves. This can be compared with the value predicted by Einstein's theory. If the expected and observed values happen to be equal, it will provide a direct validation of the existence and emission of gravitational waves.

Hulse and Taylor continued their work on the observations of the binary pulsar system and in 1978, four years after its discovery, announced that they had indeed found a decrease of the binary period which was consistent with the prediction by Einstein's theory. The general relativistic value was obtained using the measured parameters of the binary pulsar. This was the first time that the emission of gravitational waves predicted in 1916 was found to really exist. Observations accumulated over the years and reported in 2010 have shown that the observed rate of change of orbital period and the prediction of Einstein's theory agree to better than one percent. The change in orbital period changes the time taken by the pulsar to pass the perihelion in successive orbits. Because the orbit shrinks in size, the perihelion passage occurs somewhat earlier for successive orbits. The cumulative shift over many years of observation is shown in Fig. 6.3. The continuous curve shows the cumulative shift with time as predicted by Einstein's theory, while the dots are the observed values. The dots fall very closely on the predicted line, with any observed departure from the line being too small (smaller than 1%) to be seen. The emission of gravitational waves is therefore very well validated.

Hulse and Taylor were given the Nobel prize in physics in 1993. The Nobel citation said that the award was *for their discovery of a new type of pulsar, a discovery that has opened up new possibilities for the study of gravitation.*

Chapter 7
Gravitational Wave Detectors

Abstract In this chapter, we describe the laser interferometric gravitational wave detectors, like LIGO, which was used to make the first detection of gravitational waves in 2015. A gravitational wave passing through a circle of test particles will cause the shape to oscillate from a circle to an ellipse to a circle repeatedly, because of the periodic stretching and shrinking of space-time caused by the wave. As we describe, this simple fact can be exploited through a Michelson interferometer to detect the passage of gravitational wave by the changes it produces in the fringe pattern seen in the detector. The LIGO, VIRGO and other detectors are basically very large and sophisticated interferometers which are sensitive to changes in length as small as 10^{-19}m produced by gravitational waves. This sensitivity requires the use of very long detector arms in very high vacuum, very powerful lasers to produce the light beams needed by the interferometer and extreme isolation of the mirrors involved from all external vibrations. The needed sensitivity was achieved through heroic scientific and technical efforts over decades. We describe in detail the LIGO detectors in Louisiana and Washington, and mention the VIRGO detector in Italy, the GEO600 detector in Germany and the TAMA300 and KAGRA detectors in Japan. We briefly describe the forthcoming LIGO-India project.

7.1 Introduction

The very first direct detection of gravitational waves was announced on February 11, 2016. The detection was made by the *LIGO* Observatory on September 14, 2015. But the announcement had to wait until it was confirmed that the signal which had been observed was really due to the passage of a gravitational wave across the LIGO detectors, and not a spurious signal of some kind. The detection was the culmination of about four decades of effort to design and build laser interferometric detectors with sufficient sensitivity to detect very weak gravitational wave signals. The earliest attempts to detect gravitational waves were made by Professor Joseph Weber of the University of Maryland in the early 1960s, who used *bar detectors*, which came to be known as Weber bars. Weber announced in the 1960s that he had detected gravitational waves from the centre of our galaxy, but his results could not

A. Kembhavi and P. Khare, *Gravitational Waves*,
https://doi.org/10.1007/978-981-15-5709-5_7

be reproduced by several later independent efforts made using similar technology. In this chapter, we will first consider the effects produced by gravitational waves which are relevant to detectors, follow that by a brief discussion of bar detectors and then consider in greater detail the laser interferometric detectors which have proved to be so successful in finding gravitational wave sources.

7.2   Effects Produced by Gravitational Waves

The effects produced by an electromagnetic wave are easy to understand and discern. An electromagnetic wave passing in some direction exerts a force on any charged particles in its path, like electrons, and accelerates them. The accelerated particles in turn emit radiation, which results in some of the energy in the electromagnetic wave being scattered in various directions. If the wave encounters an antenna, it sets up a current in it, which can be easily detected. Modern life is closely tied to such effects produced by electromagnetic waves.

The effects of gravitational waves are more subtle and weak, and therefore far more difficult to detect. We have seen in Chap. 4 that gravitational waves are ripples in four-dimensional space-time. When a gravitational wave passes by some location, the geometry there gets affected. As a result, there is change in all distances around the location. To see the effect of such change, consider tiny particles which are arranged in a circle in space, as shown in Fig. 7.1. Now suppose a gravitational wave passes in a direction perpendicular to the circle into the plane of the paper. The change in the geometry produced by the wave is such that there is alternately expansion and contraction of distances along the horizontal and vertical directions in the plane of the paper: when there is contraction along the horizontal as shown in the figure, there is expansion along the vertical direction. As time passes, and the gravitational wave propagates forward, the directions of contraction and expansion are exchanged. For particles which are not along the horizontal or vertical directions, whether expansion or contraction takes place depends on the direction in which they are located. The net effect is that the circle of particles changes shape periodically: it alternately becomes an ellipse first stretched along the x-direction and then stretched along the y-direction, as shown in the upper part of the figure.

The periodic change of shape can occur in a somewhat different way, with the maximum stretching taking place in a direction which is inclined to the x- and y-directions, as shown in the lower part of the figure. This occurs because of difference in a property of the gravitational waves known as polarisation. If a gravitational wave approaches the circle in a direction which is not perpendicular to the plane of the circle, or there is more than one gravitational wave passing through at the same time, then the effects will be similar to those described, with the details depending on the situation.

The particles we considered above are *free particles*, in the sense that they are not under the influence of any force and are affected only by the gravitational wave passing through. The distance between the particles changes as the space expands

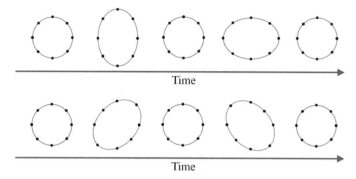

Fig. 7.1 The effect of a gravitational wave on a set of free point masses arranged in a circle. The gravitational wave travels into the plane of the paper, in a direction perpendicular to the plane. The upper and lower parts of the figure correspond to two polarisation states, as explained in the text. Image credit: Kaushal Sharma

and contracts due to the passage of the gravitational wave. But what about a bar of steel, say, which is in the path of the gravitational wave? Would the rod expand and contract too, if the effect of the gravitational wave is universal?

The situation with the steel bar is quite different from that of the particles considered above. The bar is made of particles, in the form of atoms and molecules, which are in incessant motion, and interact with each other through electrical forces. It is these forces which hold the particles together so that the bar acquires a form which is rigid on the large scale. When the bar changes its length due to the expansion and contraction of space due to passage of a gravitational wave, there is a change in the forces between the particles, and restoring forces are set up. These cause the bar to to return to its original size. The result of the two opposing effects is that the bar begins to vibrate; these vibrations continue even after the wave has passed by, until they are gradually damped out. This is like the ringing of a bell when it is struck, which persists for some time after the blow, and gradually reduces in intensity. The vibrations of the bar are of very low amplitude, i.e. they are very feeble, so it is extremely difficult to detect them. For a bar of cylindrical shape of a given size and material, there is a particular frequency, known as the *resonance frequency* which has the greatest amplitude. This frequency depends on the length of the bar, its density and on how elastic the material of the bar is.

7.3 Bar Detectors

In the 1960s, Professor Joseph Weber of the University of Maryland decided to measure the bar vibrations triggered by gravitational waves to detect their passage. For this purpose, he constructed a number of accurately machined bars of aluminium, one of which was 2 m long and 1 m in diametre, with other bars having similar

Fig. 7.2 Professor Joseph Weber and one of his bar detectors at the University of Maryland. The cylinder of aluminium and the circle of detectors around the middle can be seen. Image courtesy of University of Maryland Libraries

dimensions. The idea was that when a gravitational wave of a specific frequency passed by the bar, the bar would begin to vibrate or ring, as described in the previous section. While vibrations with various frequencies would be present, the most dominant would be the resonant or ringing frequency. For the cylinder used by Weber, this resonant frequency was close to 1660 Hz, i.e. 1660 vibrations per second. This frequency was chosen as Weber expected it would be the dominant frequency emitted by various gravitational wave sources.

The vibrations, being very feeble, are very hard to detect and Weber used piezo-electric crystals for the purpose. These crystals produce electric voltage when they are stretched or compressed. The crystals were placed on the surface in the middle of the cylinder. They responded to any vibrations in the bar by generating tiny voltages which could be measured using sensitive electronic instruments. Prof. Weber with his bar detector is shown in Fig. 7.2.

While a gravitational wave passing by the bar should set it ringing, that could also happen due to other mechanical disturbances, while electromagnetic disturbances can affect the crystals, producing wrong signals. The ground has low amplitude seismic vibrations, as well as those generated by human activity and these can induce vibrations in the bar, thus masking the vibrations due to gravitational waves. The bars were suspended from steel wires in vacuum to minimise mechanical disturbance and were also shielded from electromagnetic disturbances. A source of noise which could not be decreased was that caused by incessant random motion of the atoms of Aluminium. The greater the temperature of the bar the faster is this motion. At the room temperature that Weber's bars operated, the random motion led to irregular changes in the length of the bar to a very small extent, 10^{-16} m, which was comparable to the expansion and contraction of the bar expected from the effects of gravitational waves.

Weber

Professor Joseph Weber was an American physicist who spent much of his working life at the University of Maryland. He was in the American Navy during the Second World War, and was posted on the aircraft carrier USS Lexington, which was sunk in action with the Japanese Navy. Weber survived the experience, served on other ships and took part in the invasion of Sicily. Prof. Weber was a theoretician as well as an experimentalist, and did pioneering work in two very different fields, first masers and lasers, and then gravitational wave detection. He discovered the basic principles behind masers and lasers and gave the first ever lecture on the subject. He was jointly nominated for a Noble Prize for their discovery, but the Prize eventually went to three other physicists. Weber then turned to the possibility of detecting gravitational waves and he built a bar detector for that purpose. After a series of experiments, he announced that he had detected the waves. While his results were published in the best scientific journals, questions remained about their validity, and increasingly sophisticated bar detectors developed by other groups failed to detect gravitational waves.

Weber announced in the late 1960s that he had detected gravitational waves from the signals he observed from his bar detectors. He later moved one of his detectors from his laboratory in the University of Maryland to the Argonne National Laboratory near Chicago, about a thousand km from his laboratory. The idea was that two detectors so far away from each other could not have the same ambient *random noise*. He found many coincident detections in the two bars which he claimed were of sufficient strength to be definitively detected over the noise. He also concluded that the source of the gravitational waves was situated in the centre of our galaxy.

Weber's results could not be reproduced by other similar experiments and his analysis of the data was believed to be incorrect. The community of scientists therefore did not accept that Weber had detected gravitational waves. But Weber's work and his enthusiasm inspired many other groups to develop other versions of bar detectors, with improved ability to detect the waves. One major change was to cool the detectors to very low temperatures using liquid helium, so that the noise due to random motions of the atoms was greatly reduced. Motion sensors which were more sensitive than *piezoelectric crystals* were used, and in some cases a spherical rather than cylindrical shape was used for the detector. While the cylindrical bar was mainly sensitive to gravitational waves travelling perpendicular to the axis of the cylinder, spherical detectors were equally sensitive to waves coming from any direction. Several resonant detectors were used over three decades following Weber's experiments, but there was no detection of gravitational waves by any of these detectors. The first detection had to wait until 2015, when the Advanced LIGO detectors, which are different from the bar detectors, were finally successful.

7.4   Interferometric Detectors

These detectors are based on the Michelson interferometer, which was developed in 1881 by the American physicist A. A. Michelson. The use of interferometry for detecting gravitational waves was first proposed in 1962 by the Russian physicists Michael Gertsenshtein and Vladislav Pustovoit and later independently by Joseph Weber and Rainer Weiss in the USA. Weiss fully developed the concept, and prototype detectors were built in several places. It was decided to have large interferometers at two sites, one at Hanford, Washington State and the other at Livingston in Louisiana. The construction of the LIGO detectors was jointly taken up by the California Institute of Technology (Caltech) and the Massachusetts Institute of Technology (MIT) with funds provided by the National Science Foundation. The *initial LIGO*, or iLIGO, interferometers began gravitational wave searches, in 2002, which continued for about a decade. While no detection was made in this period, it was possible to develop deeper understanding of the advances which would be required to improve the detector's sensitivity to a level which would lead to actual detections.

The experience gained in pushing initial LIGO to achieve the sensitivity it was designed for was used in developing the Advanced LIGO, or *aLIGO*, interferometers which were ready for observations around the middle of 2015, with a sensitivity about 10 times higher than initial LIGO. The first detection was made in September 2015, shortly before the beginning of the first systematic searches by Advanced LIGO. Since 1997, the scientific and technical work related to LIGO has been carried out under the *LIGO Scientific Collaboration (LSC)* which by 2016 had about 1000 scientists working in 15 countries. From 2007, the work has been done in collaboration with the European *VIRGO* laser interferometric detector.

7.4.1 *Michelson Interferometer*

The interferometer named after him was developed by Michelson in 1881 to very accurately measure lengths and changes in length using the phenomenon of interference of light. In 1887 Michelson and Morley used the instrument to investigate the effect on the speed of light due to the passage of the Earth through the all pervading ether, which was then believed to exist, as we discussed in Sect. 2.4.

A very simple version of the Michelson interferometer is shown in Fig. 7.3. A beam of light emitted from the source at the left of the diagram travels to a mirror which is known as a beam splitter. This mirror allows half the light of the beam to pass through it, towards mirror M_1, while the other half of the light is reflected towards mirror M_2. The light striking the mirrors M_1 and M_2 is reflected back towards the beam splitter, and the combined beam then reaches a light detector at the bottom of the figure. The two paths of the beam between the beam splitter and M_1, and the beam splitter and M_2, can be viewed as two arms of the interferometer. We have seen in Sect. 2.3.1 that light is a form of electromagnetic radiation and propagates as a wave. When the two beams are combined together, the result depends on how the corresponding waves are matched. When the distance travelled by the two beams is exactly the same, the waves combine together constructively, producing a bright spot at the detector. If the distance travelled differs by half the wavelength of the light wave, then the two waves combine destructively, producing a dark spot. Depending on the exact configuration of the mirrors M_1 and M_2, the combined beam can produce concentric light and dark circles, or a pattern of light and dark straight lines at the detector. If either M_1 or M_2 are moved slightly, the distance travelled by one of the beams changes, leading to a

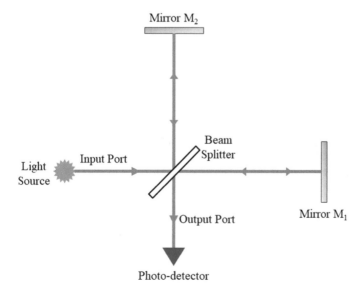

Fig. 7.3 A simple version of the Michelson interferometer. Image credit: Ajith Parameswaran

change in the pattern observed. The process of combining two wave trains of light to act constructively or destructively is known as *interference* and the pattern produced is known as an *interference pattern*.

Michelson
Albert Abraham Michelson was an American physicist who is best known for the Michelson-Morley experiment which was used to rule out the existence of the ether. Michelson was awarded a special appointment to the United States Naval Academy at Annapolis by President Ulysses S. Grant, where he later became an instructor. At Annapolis, he conducted his first experiments to measure the speed of light. He obtained his Ph.D. from Europe working under the famous physicist Hermann Helmholtz, and in 1883 was appointed a Professor of Physics at the Case School of Applied Science in Cleveland, Ohio, which is now known as the Case Western Reserve University. There Michelson worked on developing an accurate interferometer and in 1887, working with Professor Edward Morley, carried out the Michelson-Morley experiment. The experiment was designed to detect the ether, but failed to do so. The result could only be understood after the development of the special theory of relativity by Albert Einstein. For his work in optics, Michelson was awarded the Noble Prize for physics in 1907. He was the first American to receive a science Noble Prize. Working with Francis G. Pease, in 1920 Michelson used an interferometer at the Mount Wilson Observatory to measure the diametre of the super-giant star Betelgeuse. Until then, the only star for which the diametre had been measured was the Sun.

The concept of the interferometer provides a way to detect gravitational waves. We have seen above in Sect. 7.2 that when a gravitational wave passes through a circle of free particles, the circle periodically changes shape to an ellipse which is first stretched in one direction and then in the perpendicular direction. Now imagine that mirrors M_1 and M_2 are attached to particles on a circle, with a beam splitter M placed at the centre, as shown on the left in Fig. 7.4. The three mirrors M_1, M_2 and M play the same role as the mirrors in Fig. 7.3. The distances D_1 and D_2 on the left of the figure are equal. A beam of light which strikes M is split into beams which go towards M_1 and M_2 and are reflected back to M, where they combine together and reach the detector. The distance travelled by the two beams is the same and so is the time taken for the beams to arrive at the detector, and the interference is constructive so a bright spot is seen at the centre of the detector. Now suppose a gravitational wave passes through the circle of particles, so that it is stretched to form an ellipse as shown on the right of the figure. The mirror M_1 is now farther from M than it was in the absence of the wave, while M_2 has moved closer to M, so D_2 is smaller than D_1. The distances travelled by the two beams are therefore different and for a certain difference in the path length, the interference is destructive and a dark spot is seen. As the mirrors move back and forth the interference pattern

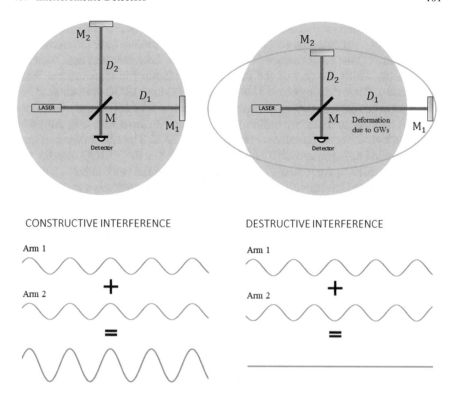

Fig. 7.4 The changing distance of mirrors on the circumference of a circle due to the passage of a gravitational wave. On the left are shown the mirrors in their normal state, when they are on the circle. The mirrors M_1 and M_2 are at equal distance from the mirror M. As explained in the text, there is constructive interference of the two beams, which is shown schematically in the lower part of the diagram. On the right the configuration is shown when a gravitational wave is passing through. M_1 has moved away from the centre of the circle, while M_2 has moved towards the centre. The mirrors now are on an ellipse. The change in distance of the two mirrors from the centre is such that the interference is destructive, as shown schematically in the lower part. Image credit: Kaushal Sharma and Ajith Parameswaran

changes periodically, which can be detected by the changing intensity of the spot. The changing interference pattern becomes a signature of the passage of a gravitational wave. We will see below how this idea of using a Michelson interferometer to detect gravitational waves is put into practice in the laser interferometric gravitational wave detectors like LIGO and VIRGO.

7.5   Laser Interferometric Detectors

These detectors are based on the Michelson interferometer and use a laser as the source of the beam. They have arms a few kilometres long as that helps in increasing the sensitivity of the detector, as explained in Sect. 7.5.5. Even longer arms could not be considered when the detectors were planned, because of practical limitations like availability of a suitable site, cost and technology. The present laser interferometric detectors are (1) the LIGO detectors near Livingston, Louisiana and near Hanford, Washington, which are part of the the Laser Interferometric Gravitational Wave Observatory. These detectors have 4 km long arms; (2) the VIRGO detector near Pisa with 3 km arms; (3) the GEO600 detector near Hanover with 600 m arms and (4) the KAGRA underground detector at Kamioka in Japan with three 3 km long arms with mirrors which are cryogenically cooled to about −23 K. The LIGO detectors have been upgraded to Advanced LIGO (aLIGO) in 2015, and the VIRGO detector was upgraded in 2017 to Advanced VIRGO. Completion of the work to build KAGRA was celebrated on October 4, 2019. The GEO600 detector is presently used to develop and establish technologies which are later used in the more sensitive detectors for actual observations. The TAMA300 detector near Tokyo with 300 m long arms has been decommissioned some years ago. We will describe below the aLIGO detectors, which have been successful in detecting gravitational waves. The other detectors are similarly constructed, with differences in detail.

7.5.1 The Advanced LIGO Detectors

Translating from laboratory scale interferometers to the LIGO scale with each arm being 4 km long is obviously rather difficult. The difficulty is greatly compounded by the accuracy that is required for the detection of gravitational waves. In the laboratory, an interferometer is used, for example, to measure the wavelength of light. This requires an accuracy of about 10^{-12} m in very precise measurements. But the displacement caused by a gravitational wave in the LIGO mirrors can be as small as 10^{-19} m, which is about 1/8500 of the radius of a proton. This extremely small displacement has to be observed against a background of much larger displacements caused by various effects.

The required sensitivity is reached by adapting a series of technical innovations designed to eliminate or minimise spurious vibrations, which we can term as *noise*. Some of the measures are (1) using a powerful laser as the source of the light beams; (2) using extra mirrors to increase the intensity of light and the distance traversed by the beam through repeated reflections; (3) placing the entire interferometer in a high vacuum enclosure and (4) using very sophisticated suspensions for holding the mirrors. The details that we describe below are for aLIGO, but similar systems are used in the other detectors.

7.5.2 Laser

Michelson interferometers originally used a conventional source of light, which could be white light, or light of a specific color, i.e. of a fixed wavelength. For such a conventional light source it is difficult to maintain accurately a given wavelength, which is required to produce a fixed interference pattern. If the wavelength changes, so does the pattern, which can be mistakenly assumed to be due to the passage of a gravitational wave. Other properties required of the light source are coherence and high intensity. In a coherent source, the relationship between the wave trains in the two beams remains constant with time, which again is required for getting a steady interference pattern. High light intensity is needed to produce good contrast between the light and dark areas of the interference pattern, but there is a more subtle effect. This can be understood by considering the light beams to be made up of photons, which are the particles of light in quantum theory. The number of photons in a cross section of the beam fluctuates randomly, leading to fluctuations in the interference pattern. Such effects which arise due to the discrete nature of photons are known as *quantum noise* or *photon shot noise*. The effect of the fluctuations as a fraction of the intensity is reduced as the intensity becomes larger, so that the interference pattern appears steady and sharp and accurate measurements can be made on it. Quantum noise also results in slight fluctuation of the pressure exerted by the laser light on the mirrors of the interferometer, leading to very small vibrations.

A laser beam is ideally suited to satisfy the above requirements. The laser provides a very intense, steady and non-diverging beam of light at a precisely determined wavelength. The complex laser device used in the LIGO detector produces a laser beam at a wavelength of 1064 nm, with power of 180 W. It is one of the most steady and powerful lasers operating at the chosen wavelength, which is in the near-infrared part of the spectrum. The laser is stabilised to a level of one part in ten billion in intensity and one part in a billion in frequency. The power produced by the laser needs further great amplification to serve the needs of the detector, which is achieved by using a partly reflective mirror, called a power recycling mirror, placed between the laser and the beam splitter for recycling the light. Most of the laser light which is reflected by M_1 and M_2 comes to the recycling mirror, rather than to the detector. The light is then reflected back to the beam splitter and from there again to M_1 and M_2. Such repeated recycling greatly increases the laser power in the beam, which can reach a design level of 750 kW, which is far higher than the input laser power of 180 W. In the first observing run of aLIGO during which the first gravitational wave detection was made, only 20 W of laser power was injected into the interferometer which resulted in 100 kW of power circulating in each arm.

The length of each arm of LIGO, from the beam splitter to the reflecting mirror at the end of the arm is 4 km. The longer this length, the higher would be the sensitivity of the detector, as explained in Sect. 7.5.5. A much longer length is effectively achieved by placing an additional mirror, called a signal recycling mirror, in each arm exactly 4 km for the mirror at the end of the arm. The 4 km long space is known as a Fabry-Perot cavity. The beam in each cavity is reflected back and forth about 300 times, effectively increasing the arm length to about 1200 km, and also increasing the number of photons in the arms which helps in reducing noise as explained above.

A schematic of the aLIGO interferometer is shown in Fig. 7.5. Mirrors used for various functions, the test masses, laser and detectors are shown. The laser power in different parts of the interferometer is indicated. The input mode cleaner controls

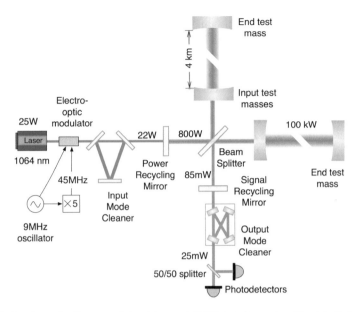

Fig. 7.5 A schematic diagram of the Advanced LIGO interferometer. The labelled parts are explained in the text. Image credit: D. V. Martynov et al. Phys.Rev.D **93**, 112004, 2016

some properties of the laser, while the output mode cleaner helps with the readout of the signal. The real LIGO interferometer is of course much more complex than the simple version shown, with multiple mirrors and other optical devices needed for each function we have mentioned. To maintain optimal operation, all the optical devices have to be accurately aligned and located, with precision better than 10^{-11} cm (one ten trillionth of a cm!).

The arrangement of the optics is such that when the arm lengths are equal, a dark spot is produced at the detector. As the arm lengths change, the interference pattern changes, and the spot gets brighter. Conventionally, equal arm lengths would lead to constructive interference and a bright spot would be seen at the detector. But in the LIGO interferometer, a dark spot is chosen by adjusting the optics, because it is easier to judge the brightening of a dark spot, rather than the dimming of a bright spot.

7.5.3   Mirrors

The mirrors in the LIGO interferometer have a double role to play: (1) they allow the laser beam to be split into two parts, one going along each arm, and then to reflect them back to be combined into a single beam and (2) they act as free particles, which means that there should be no net force, other than the gravitational force of the gravitational wave, which should be acting on them. For that, the end mirrors cannot be fixed, and have to be suspended from above. For suspended mirrors, in the vertical direction, the downwards gravitational force of the Earth is balanced by the upwards tension in the suspension. The mirrors can move freely in the horizontal direction and the effect of a gravitational wave would be to change the horizontal distance between M and M_1, and M and M_2. It is necessary of course to minimise the effects of any other forces which would result in the unwanted horizontal movement of the mirrors, which is considered to be noise, since it can conceal the tiny effect due to gravitational waves. That requires advanced mirror isolation technology, which has evolved over the years from iLIGO to aLIGO.

The mirrors which constitute the free particles have a diametre of 34 cm, thickness 20 cm and weigh 40 kg each, the large mass helping to hold them still. They are made of extremely pure and homogeneous fused silica. The mirrors have very precisely defined shapes which are accurate to better than 10^{-10} cm, which is far greater precision than is required for mirrors used in large optical telescopes. The surfaces of the mirrors are so highly reflective that they absorb only about one out of 2.3 million photons which reach the mirrors. Any absorbed photons lead to heating of the mirror. Due to the very high intensity of the light, in spite of the small absorption, the heat produced is sufficient to produce a slight distortion of the mirror surface. To avoid that, a system is used which monitors the shape of the surface and makes the necessary corrections. The suspension system of a mirror in aLIGO is shown in Fig. 7.6. This system can reduce the noise due to vibrations of the ground to a billionth of its level. Further reduction to a thousandth of this corrected level is obtained by using a seismic isolation platform which can estimate the noise and correct for it.

In spite of the advanced technology used for noise reduction, there is residual noise which can cause small displacements in the mirror. First, there is *seismic noise* due to ground vibrations or seismic activity caused by human beings, winds, tidal motions in the ground caused by the Sun and the Moon similar to tides in the oceans, and ocean waves which can cause disturbance even when the coastline is far away. The seismic disturbances cause small changes in the density of the ground, which in turn causes small changes in the local gravitational field, leading to disturbance of the mirror which changes with time. This is known as *Newtonian noise* or *gravity gradient noise*. Disturbances can also be caused due to slight heating by the laser beams of the mirrors and their suspensions and mechanical loss of the reflective coatings of the mirrors. This affects the random motion of the particles in them, leading to *thermal noise* or *Brownian noise*. The effect of these sources of noise on the sensitivity of the aLIGO detector is described in Sect. 7.5.5.

7.5.4 The Vacuum System

The entire LIGO interferometer is operated in a high vacuum. The laser beams pass through two steel tubes each of which is 4 km long and the mirrors are enclosed by end stations. The whole evacuated assembly has a volume of about 10 million litres.

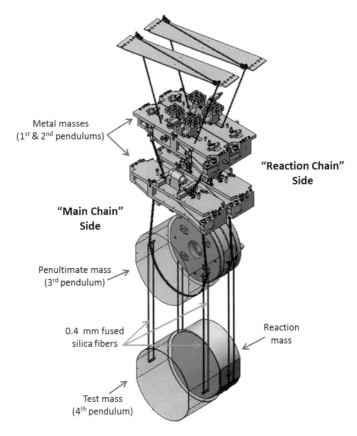

Fig. 7.6 The suspension system in aLIGO used to hold the mirrors which act as free particles. Image credit: CALTECH/MIT/LIGO Laboratory

The only bigger evacuated volume than LIGO is the Large Hadron Collider in CERN in Switzerland, which was used to discover the Higgs Boson. The vacuum is so good that the pressure of the residual gases in the volume is only a trillionth of the air pressure at sea level.

The high vacuum is required for various reasons. The molecules of air are in incessant motion due to their heat energy. The molecules collide with the mirrors to produce tiny motions in them. If the pressure is not low enough the jitter produced in the mirrors would produce noise at an unacceptable level. When light propagates through air, its direction of travel can be slightly altered depending on the density of the air. There are always small disturbances in air which can cause the beam to randomly deviate from its straight path, which changes slightly the length travelled by the beam, adversely affecting the interference pattern. The air can contain dust particles which scatter the beam from its straight line path. These effects are all very small and would not matter at all in other situations. But because of the very small changes produced by gravitational waves the effects can become important in comparison in the present case. The very high vacuum is needed to reduce such effects to an acceptable level.

7.5.5 Sensitivity of the Advanced LIGO Detector

The Advanced LIGO detectors, which were commissioned in 2015, have many improvements over the initial LIGO detectors. The improvements were made taking into account the experience gained from iLIGO and the many improvements in technology which have become possible over the twenty years which separate the two detectors. For example, the earlier laser power was 10 Watts while the laser in aLIGO is of 180 Watts; the aLIGO mirror weighs 40 kg while against the 11 kg for the earlier mirrors. iLIGO could detect gravitational waves extending in frequency from 40 to 100 Hz, while aLIGO can detect frequencies over a broader range from 10 to 1000 Hz. This extension of the lower frequency range enables the detection of some kinds of sources which were not earlier detectable. Overall, aLIGO is about 10 times more sensitive than iLIGO. This allows aLIGO to detect a given source over 10 times the distance over which iLIGO could have detected it. aLIGO therefore probes 1000 times the volume that was available to the earlier detector, since volume increases as the cube of the distance. This greatly enhances the number of sources one can expect to detect with aLIGO. The greater sensitivity also makes possible more detailed study of the detected sources.

The sensitivity of the aLIGO detector to gravitational waves and to noise in the system as a function of frequency is shown in Fig. 7.7. The frequency at which the measurement is made is shown on the horizontal axis, while a quantity known as the *strain noise* is shown on the vertical axis. This quantity is indicative of the ratio of the net displacement produced in the two arms to the total distance traversed by the beam in the multiple reflections discussed above. The curve in dark red shows the strain noise for the aLIGO detector at Hanford during the first observing run O1, which is described in Chap. 8. The curve in light red is for the detector at Livingston. The near coincidence of the two curves shows that the sensitivity of the two detectors has nearly the same dependence on the frequency.

The *strain* produced in the arms increases with the amplitude of the gravitational wave detected. Since the strain is the ratio of the displacement produced to the length of the arms, it follows that as arm length is increased, the displacement would also increase. Increased displacement would make the detector less prone to the effects of noise, so arm lengths greater than those of aLIGO would be an advantage. We will see in Chap. 9 how greater arm lengths may be achieved in future projects.

A gravitational wave source can be detected by aLIGO only if the strain produced by the source is above the curves. How much confidence we can place in the detection depends upon how much above the curve the source is located. Therefore, the lower a curve is on the diagram, the better the detector would be for observing faint sources. Using this criterion, it is seen that the aLIGO detectors are most sensitive in the

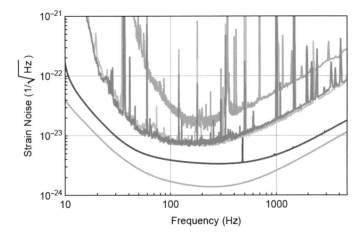

Fig. 7.7 The sensitivity of the aLIGO and iLIGO detectors as a function of frequency. See text for a description of the figure. Image credit: B. P. Abbott et al. PRL **116**, 131103, 2016

frequency range of 100–300 Hz. For lower frequencies, the red curves show greater strain noise, which is dominated by seismic, Newtonian and thermal noise, as well as instrumental noise. The increase in strain noise above 100 Hz is due to quantum noise, which dominates in this region. Some of the noise sources in the frequency range of 20–100 Hz are not identified at the time of writing. The strong lines at various points along the red and green curves are due to specific contributions to the noise from the suspensions, from AC power lines and lines used for calibration purposes. The lines are accounted for in the analysis of the data.

The green curve in Fig. 7.7 is the strain noise for the iLIGO detectors during their last science run S6. In the band 100–300 Hz, where the detector sensitivity is the best, it is seen that the strain noise in aLIGO is about 3–4 times lower than that for iLIGO. At 50 Hz, the noise in aLIGO is about 100 times lower. It is the lower noise which has made aLIGO so successful in detecting gravitational wave sources. The aLIGO sensitivity will be improved further during 2020-21, and the design strain noise which is expected to be achieved is shown by the curve in blue. The curve in cyan is for an upgraded *LIGO A+* detector planned for the future (see Sect. 7.5.6).

An aerial view of the LIGO facility at Hanford in Washington State and a close-up of its Northern arm are shown in Figs. 7.8 and 7.9 respectively.

7.5.6   Beyond Advanced LIGO

The Advanced LIGO detector will be upgraded to LIGO A+ over a two year period beginning sometime in early 2023. The upgrade will involve building a 300 m cavity for implementing a technique known as frequency-dependent squeezing. This will

Fig. 7.8 An aerial view of the LIGO facility in Hanford, Washington State, USA. The two arms of the detector can be seen. Image credit: CALTECH/MIT/LIGO Laboratory

Fig. 7.9 A close view of a portion of the northern arm of the LIGO Hanford detector. The path of the laser beam is covered with a steel pipe which has a very high vacuum, which is in turn covered with a protective concrete layer which is seen in the image. Image credit: From Wikipedia, under a Creative Commons Licence

help reduce quantum noise through reduction in the photon shot noise at frequencies above 500 Hz as well as reduced fluctuations in the pressure of the photons on the mirrors at lower frequencies. Other improvements will include reduced Newtonian noise, a larger diametre beam splitter, test masses of 100 kg with improved coating which will lead to reduced thermal noise, laser power increased to 200 W and improved techniques for readout. The improvement in sensitivity brought about relative to the sensitivity of aLIGO is shown in Fig. 7.7. The implication for observations is mentioned in Sect. 8.6. The Advanced VIRGO detector will also be similarly upgraded.

LIGO A+ will be followed by the *LIGO Voyager*, which is being designed to have the ultimate sensitivity which can be reached with the existing LIGO infrastructure. LIGO Voyager will have test masses weighing 200 kg made of silicon, with amorphous silicon coatings. The mirrors will be cryogenic, operating at 123 K. A laser operating at about 2 micrometres will be needed to suit the change in the mirror material. The laser power circulating in the arm will be 200 MW. When the design sensitivity will be reached, LIGO Voyager is expected to detect merging compact binary sources at about a hundred times the rate reached by Advanced LIGO.

7.6   The LIGO-India Project

This is a joint project between a consortium of institutions in India and LIGO Labs in the USA. The project is for installing an Advanced LIGO detector in a facility created in India for the purpose. The detector will be provided by LIGO Labs, while the large vacuum facility needed for installing the detectors and all other elements of the gravitational wave observatory will be developed and operated by the Indian side. The detector in India will substantially improve the localisation in the sky of detected gravitational wave sources, making it much easier to detect their electromagnetic counterparts. There will also be other advantages, as explained below.

Initial LIGO consisted of three gravitational wave detectors, with one installed at Livingstone and two installed at Hanford. The detector in Livingstone and one of the detectors in Hanford had the 4 km long beam lengths we have described above. The second detector in Hanford, which was located in the same arms as the first one, had one path of length 4 km and the other 2 km. When aLIGO was designed, it was decided that all three detectors would have paths with equal length of 4 km, with one detector in Livingston and two in Hanford as before.

It was realised some years ago by LIGO Laboratory that there would be great advantage if one of the two aLIGO detectors which were to be deployed at Hanford was instead transferred to a location at a great distance from the USA. A single LIGO detector operating by itself is rather poor at determining the direction in the sky where a given detected source of gravitational waves is located. The detectors in Hanford and Livingston are at a distance of about 3000 km from each other and when these detectors observe in tandem, there is some improvement in determining the direction. But to get better results, it is necessary to locate the third detector as far away as

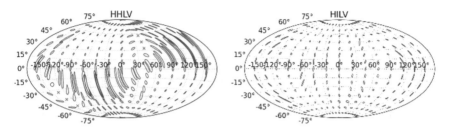

Fig. 7.10 Maps of the sky comparing how accurately the positions of compact binary merger sources can be located in the sky. The figure on the left shows the contours in which a source would be located, for a configuration consisting of three aLIGO detectors in the USA and the Advanced VIRGO detector in Italy. The figure on the right shows contours for a configuration with two aLIGO detectors in the USA, the Advanced VIRGO detector and one aLIGO detector in India. The improvement in source localisation is obvious. Image credit: B. S. Sathyaprakash et al., LIGO Technical Note LIGO-T1200219-v1 (2012)

possible from the two detectors in the USA. The three aLIGO detectors observing together would constitute a giant triangle and together with Advanced VIRGO provide the best directionality. The four detectors would also provide improved overall coverage of the sky, more precise measurements of various parameters of interest of the detected sources, and the longest period over which at least three detectors would be available for triangulation, i.e. for determining the direction of the detected source in the sky.

Around the year 2010, LIGO Laboratory began discussions with gravitational wave physicists in India to consider the possibility of installing an aLIGO detector at a suitable location in the country. With a detector in India, the longest baseline, i.e. the longest distance between detectors would be such that a gravitational wave, travelling at the speed of light, would take about 39 ms to go from one to the other. It takes only a maximum of 10 ms for a gravitational wave to travel between Hanford and Livingston. The longer time made possible by the longer baseline very significantly improves the localisation of gravitational wave sources in the sky, as shown in Fig. 7.10.

India has a rich tradition of research in gravitational theory and gravitational wave physics and astronomy. There is a significant number of people engaged in developing gravitation wave data analysis algorithms and in the actual data analysis. Sensitive gravitation related experiments have been done at the Tata Institute of Fundamental Research, Mumbai over the years. A number of bright young persons who obtained their Ph.D. in India or abroad in all these and other related areas have been appointed as faculty in various research institutions and university departments in India. Many of these researchers have come together under the banner of *Indian Initiatives in Gravitational Wave Observations (IndIGO)*, which was established in 2009 as a consortium of interested researchers to work together to participate in developing experimental facilities for gravitational wave detection, and in their observations and data analysis. There are many research establishments in India with teams of scientists and engineers engaged in experimental work and experience in executing large projects. It was therefore possible that a LIGO facility could be created in the country through a close collaboration between those interested in gravitational wave physics and astronomy, and experimentalists. This would be a great opportunity to bring very advanced technology to the country, and for Indian scientists to directly participate in the making of the most exciting scientific discoveries of our time.

Since an Advanced LIGO detector in India would be a paradigm changing development, IndIGO and the Inter-University Centre for Astronomy and Astrophysics (IUCAA) in Pune took the initiative to have discussions with LIGO Laboratory to prepare the *LIGO-India project*. After detailed consideration, LIGO-India emerged as a national project funded by the Government of India, with the Institute of Plasma Research (IPR) in Gandhinagar, the Raja Ramanna Centre for Advanced Technology (RRCAT) in Indore and IUCAA, as the three lead institutions for executing the project. Later the Directorate of Construction, Services and Estate Management (DCSEM) of Mumbai joined the project as another lead institution. Contributions are being made to the project by many other institutes, university departments and individual researchers working in various places in India. The project is being coordinated by the Department of Atomic Energy (DAE) and the Department of Science and Technology.

LIGO Laboratory will provide to LIGO-India, free of cost, all the components needed for making one complete Advanced LIGO detector. Some of these components have been supplied to LIGO Labs by institutions in Germany, Britain and Australia. The detector hardware will include a high power continuous wave laser, all the optical elements including the mirror and other hardware. In addition to the detector, an advanced vibration isolation system will be provided. LIGO Labs will also supply the needed control systems and electronics, data analysis algorithms and protocols used in the aLIGO detectors and installation tools and fixtures. The components to be supplied by LIGO Labs are of the most advanced kind in their class, developed through years of research and experimentation.

LIGO-India will construct the $4\,km \times 4\,km$ vacuum system and all the complex infrastructure required for a gravitational wave observatory and develop the computing facilities needed for data management and analysis. LIGO-India will train and provide the human resources needed for the observatory, ranging from laboratory technicians to expert engineers to researchers. After completion of the project, the observatory will be managed by a suitable organisation.

LIGO-India will be located at a site close to the town of Hingoli in the state of Maharashtra in India. It is expected that the project will be completed by 2026.

Chapter 8
Gravitational Wave Detections

Abstract We begin this chapter by describing the first detection of gravitational waves made in September 2015 by the two LIGO detectors and follow that with a summary of the properties of some more binary black hole sources detected by LIGO and VIRGO in the subsequent observations. We then describe the spectacular neutron star binary merger which provided an opportunity to test some long-held astrophysical beliefs. The signal from the first detection lasted for just about 0.2 s. But the development of the signal over this period was very close to the shape expected from Einstein's theory for the merger of two black holes spiralling into each other. Careful analysis of the signal showed that masses of the two black holes were much greater than black hole masses expected from the observation of X-ray binary sources. The very first detection thus led to the identification of an unexpected class of black holes and binary systems, and to the 2018 Noble Prize in Physics for Reiner Weiss, and Barry Barish and Kip Thorne. The neutron star merger which was detected by LIGO and VIRGO was observed over many wavelengths of the electromagnetic spectrum, leading to a great deal of theoretical work and modelling, amply fulfilling the promise expected from multi-messenger astronomy. The many gravitational wave detections now being regularly made have helped establish gravitational wave astronomy as a very important and productive domain.

8.1 Introduction

The first confirmed detection of gravitational waves was made by the Advanced LIGO detectors on 14 September 2015. The analysis of the data to confirm that the detected signal was indeed due to a gravitational wave and not an artefact took several months, and the announcement of the detection was finally made on 11 February 2016. By coincidence, the detection was made exactly a century after the formulation of the general theory of relativity by Albert Einstein in 1915, and the announcement was made exactly a century after Einstein first predicted the existence of gravitational waves in 1916! The detailed form of the observed signal is exactly as predicted by general relativity for a pair of black holes which are in orbit around each other, move closer together because of the emission of gravitational waves, and finally merge to

A. Kembhavi and P. Khare, *Gravitational Waves*,
https://doi.org/10.1007/978-981-15-5709-5_8

form a single black hole. The discovery established the existence of black holes with mass much greater than expected, binary black holes, gravitational waves and the correctness of Einstein's theory in describing the system. It, therefore, ranks as one of the greatest discoveries in the history of physics and astronomy, which yet again demonstrates how the building of novel, large telescopes and detectors always leads to unexpected discoveries and novel phenomena.

8.2   The First Detection GW150914

We have described in Sect. 7.5 the LIGO detector, and how this was upgraded to the Advanced LIGO detector, which is up to 10 times more sensitive than the initial LIGO detectors. aLIGO detectors were installed in the LIGO observatories at Livingston, Louisiana and Hanford, Washington State in the USA. Engineering runs, during which the systems were being validated, had begun by August 2015. An announcement for the start of regular observations was to be made in September that year.

A signal, which could be a burst of gravitational waves, was observed by the detectors in the morning of 14 September 2015 at 09:50:45 UTC, where UTC is Coordinated Universal Time which in practice agrees with the better known Greenwich Mean Time (GMT). The signal was first detected at Livingston and 7 ms (7 thousands of a second) later in the detector at Hanford. The distance between the two detectors is such that a gravitational wave travelling between them along a path as shown in Fig. 8.1 would take about 10 ms, given that gravitational waves travel with the speed of light. If the wave travelled along a path nearly perpendicular to the line, it would reach the two detectors nearly simultaneously. The fact that there was a 7 ms delay between Livingston and Hanford provides constraints on the direction in which the gravitational wave could have approached the Earth and reached the two detectors as observed.

But is the signal really a burst of gravitational waves? That seemed very likely since it was detected in both the detectors within a time span which was consistent with the distance between them. But there are many other effects which could lead

Fig. 8.1 The time taken by a gravitational wave to travel between the detectors at Hanford (H1) and Livingston (L1). The direction of the detector arms at Hanford and Livingston are indicated by the figures near H1 and L1, respectively. Image credit: B. P. Abbott et al. PRL **116**, 061102, 2016

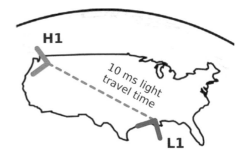

to spurious signals which mimic gravitational waves, and these have to be carefully eliminated if the signal is to be accepted as a real gravitational wave. At each site, there are many environmental sensors which record various types of disturbances which can lead to spurious signals. It was established that any external disturbance which was large enough to produce the observed signal, would have been easily detected by the environmental sensors, but no such record was seen in them. Moreover, any such disturbances would have to independently occur at the two widely separated sites within 7 ms, and it can be shown using statistics that the likelihood of that occurring is extremely low. There is the possibility that there are disturbances which are correlated over a long distance in such a way that they produce nearly simultaneous spurious signals in the two detectors. The possibility of such effects was eliminated after a careful study. The analysis therefore led to the result that the signal is real with a probability greater than 99.9999%, which means that there is less than one in a million chance that the signal is spurious. The gravitational wave detection is labelled as GW150914, with the numbers indicating the discovery date in the order year-month-day.

8.2.1 The Nature of GW150914

The gravitational wave which has been detected must have been emitted by an astronomical source, and we expect that it is one of the possible types of sources of gravitational waves that we have discussed in Sect. 5.5. Its nature can be established through detailed mathematical analysis of the signal, but much is already evident from the form of the detected signal, which is shown in Fig. 8.2.

The upper panel of the Fig. 8.2 shows the event as detected at Hanford. In the lower panel of the figure, the event as detected at Livingston is shown, together with the detection at Hanford, so that the two signals can be compared. The time from the start of the signal at 09:50:45 UTC is shown on the horizontal axis, while the vertical axis shows the strain, that is the change in length of the detector arm divided by its steady length in the absence of the wave. As the wave passes the detector, each arm periodically becomes longer and shorter than its steady length, so the strain goes from being positive to negative to positive again repeatedly. Each such oscillation is known as a cycle. The time covered by the signal is about 0.2 s, during which the strain goes through about 8 cycles, with increasing height of the peak at each cycle. The time interval between successive peaks in the signal becomes shorter, which means that the gravitational wave frequency of the signal increases for each cycle. Such a signal is known as a *chirp*. The observed amplitude of the signal, that is the height reached by the peaks, is maximum at a gravitational wave frequency of about 150 Hz, beyond which the amplitude decreases over successive cycles, which is known as the *ringdown*.

From the nature of the signal with its increasing frequency and amplitude, it follows that the source is very likely to be a binary system, with the two components rapidly going around each other. Such a binary emits gravitational waves and thus

Fig. 8.2 The upper panel
shows the signal from
GW150914 as detected at
Hanford. The lower panel
shows the signal detected at
Livingston, along with the
signal detected at Hanford in
inverted form for
comparison. Image credit: F.
J. Raab and D. H. Reitze,
Current Science, **113**, 657,
2017

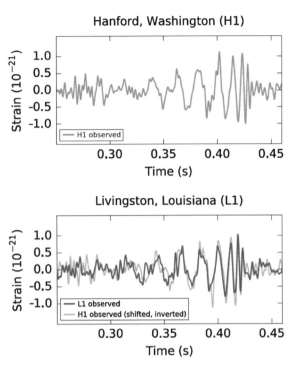

loses energy continuously, which causes the two components to spiral towards each other. This leads to decrease in the distance between them, increase in the frequency with which they go around each other, which is known as the orbital frequency, and consequent increase in the velocity of the orbital motion. The frequency of the gravitational waves emitted by the binary system is primarily twice the orbital frequency. The highest observed gravitational wave frequency of the signal reached is about 150 Hz which implies that at that point of the orbit the components should be going around each other with a frequency of 75 Hz, that is 75 times a second. The strain produced in the detector is maximum at this frequency.

From the frequency of the gravitational wave emitted from some point of the orbit, and the rate at which that frequency increases, it is possible to estimate a quantity known as the *chirp mass*. This turns out to be about 30 Solar masses. For a binary consisting of two black holes, it can then be deduced that the total mass of the two black holes should be at least about 70 Solar masses. The sum of the Schwarzschild radii of the two components must then be greater than about 210 km. This is because the Schwarzschild radius of each black hole is proportional to its mass, and the radius corresponding to one Solar mass is 3 km. Because the two components are going around each other 75 times per second when the maximum strain is reached, assuming the masses are equal, the distance between them must be only about 350 km. The components, therefore, have to be very compact. If they were extended, they would smash into each other long before the distance between them reduced to

350 km, and in that case, the gravitational wave frequency reaching 150 Hz would not have been observed, as the binary would have been destroyed. Soon after the frequency at peak amplitude is reached, the two components merge into each other, and the product of the merger is a single massive spinning black hole. The decreasing amplitude and frequency observed beyond the peak are due to the ringdown of this black hole.

Could the two components of the binary be neutron stars, which have a radius of only about 10 km and so could approach each other to the close distance inferred above? We have seen in Sect. 5.4.2.1 that a neutron star cannot have a mass greater than about three Solar masses. So the components cannot both be neutron stars, since their combined mass would not exceed six Solar masses, which is too small for the observed chirp mass. A binary consisting of a large mass black hole and a neutron star is also ruled out: in this case, it can be deduced from the chirp mass that the black hole mass would be as large as 1000 Solar masses. In such a case the Schwarzschild radius of the black hole would be about 3000 km, while at the observed maximum orbital frequency of 75 Hz, the separation would be about 850 km, which is much smaller than the Schwarzschild radius. Therefore the neutron star would be swallowed by the black hole and the binary would no longer exist well before the observed maximum frequency of 75 Hz is reached. It follows that both the compact objects in the binary should be black holes.

The above arguments which have led us to conclude that the components of the binary must be black holes are appealing and point us to the truth, but they do not let us estimate with certainty the masses of the two components and other parameters which define the binary, and the properties of the merged object. For that it is necessary to analyse the system using the general theory of relativity. The analysis involves mathematical calculations as well as the use of numerical methods. Such analysis is used to predict the shape of the signal which would be observed from black hole binaries with a wide range of properties. By comparing the predicted shape with the observed signal, the parameters which lead to the best agreement can be determined. The result is that the binary system which emitted the gravitational wave signal was at a distance of about 1.34 billion light years from us. The mass of the two black holes was 36 Solar masses and 29 Solar masses, respectively, so that the combined mass before the merger was 65 Solar masses. As described above, the binary contracts in size as the two black holes rapidly spiral-in. The merger leads to the formation of a rapidly spinning black hole with a mass of 62 Solar masses. The mass of the final black hole is, therefore, less than the combined mass of the two black holes in the binary by 3 Solar masses.

The distance of 1.34 light years of the binary from us is determined directly from the gravitational wave signal. This is possible because the amplitude of the gravitational wave is inversely related to the distance: if, for example, the distance is increased to ten times its original value, the amplitude falls to a tenth of its value at the shorter distance. The distance, therefore, emerges from the analysis of the gravitational wave signal, along with other quantities like the mass of each black hole. This is an important outcome. In astronomy, distances are generally determined using visible light. The determination depends on a long chain of distance measurements of objects which are increasingly farther away, known as the cosmic distance ladder. This has various uncertainties and having an independent distance measurement using gravitational waves is very useful. In the case of the black hole binary of course, no electromagnetic waves are emitted, so the usual method for measuring distances would simply not work. In the binary neutron star merger discussed in Sect. 8.5, gravitational waves as well as electromagnetic waves have been observed, so the results of the two methods for measuring distances can be compared.

What happened to the difference of about 3 Solar masses? According to the special theory of relativity, mass and energy are equivalent, so the missing mass must have been emitted from the system as energy. Since the components of the binary are black holes, no electromagnetic energy can be emitted, and the only form of emission possible is gravitational waves. It can in fact be shown from the theory of such sources that the total amount of gravitational wave energy emitted is equal to the energy corresponding to three Solar masses. This is exactly equal to the mass missing after the formation of the single black hole, as mentioned above. The gravitational energy was emitted in a fraction of a second, and at its peak the rate of the emission was equivalent to converting 200 times the mass of the Sun to energy in a second.

After the peak frequency is reached the two black holes merge together and the waveform enters the ringdown phase through which the merged object settles down to a black hole which has only two properties. These are the mass, which for GW150914 is 62 Solar masses, and spin, which is about 100 rotations per second. Such an object is known as a *Kerr-Newman black hole*. The observed ringdown phase agrees perfectly with the form predicted from a theoretical calculation by C. V. Vishveshwara in 1970. The calculated waveform emitted by the binary is shown in the upper panel of Fig. 8.3, while the decreasing separation of the two components, and the increasing velocity are shown in the lower panel of the figure. At their closest distance before merger, the black holes are moving relative to each other with a very high velocity of 0.6 times the velocity of light.

It would be very useful to fix precisely the location of the binary in the sky, as then any electromagnetic emission from the coalescing object could be detected. Such emission is not expected when both components of the binary are black holes. But even in that case, it would be helpful to detect any other cosmic source that the binary is associated with. For fixing the location precisely, observations from multiple locations would be needed. The source could have been observed by the VIRGO detector in Italy, but that detector was shutdown for upgradation. The GEO 600 detector was in operation but was not observing, and it would not have enough sensitivity to detect GW150914. So only the two LIGO detectors observed the source and therefore its location remains uncertain in an area of 600 square degrees on the sky.

GW150914 is the first binary black hole to be discovered. Such an object emits gravitational waves but not energy in any other form, such as electromagnetic waves. It would therefore not be detectable by any telescope except a gravitational wave detector. It is remarkable that the first source to be detected by Advanced LIGO is of such a nature. The time, effort and funds spent in the development of aLIGO are therefore completely justified, and we can expect many such surprises in the future.

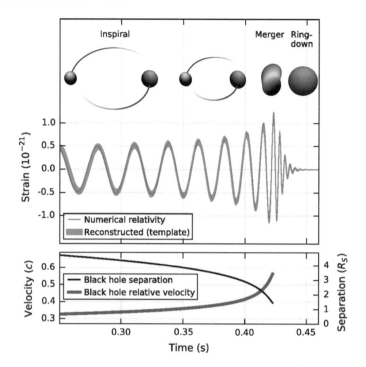

Fig. 8.3 GW150914. Upper panel: The waveform as estimated from theory is shown, together with numerically calculated models of the black holes at various stages of the eventual merger. Lower panel: the decrease in separation as the components spiral-in, and the consequent increase in their relative velocity are shown. Image credit: B. P. Abbott et al. PRL **116**, 061102, 2016

The discovery is of such great importance that the 2017 Noble prize in physics was awarded jointly to the American physicists Reiner Weiss, Barry C. Barish and Kip S. Thorne "for their decisive contribution to the LIGO detector and the observation of gravitational waves".

Physics Nobel 2017
The Noble Prize in Physics for 2017 was awarded to Rainer Weiss, Barry C. Barish and Kip S. Thorne, with half of the prize going to Professor Weiss, and the other half jointly to Professors Barish and Thorne. The prize was given for the decisive contributions that the scientists made to conceptualising and setting up of the LIGO detector and the eventual detection of gravitational waves. Reiner Weiss is a Professor Emeritus at the Massachusetts Institute of Technology in the USA. He has made pioneering measurements of the spectrum of the cosmic microwave background radiation and made major contributions to the development of the LIGO detector. Barry C. Barish is the

Linde Professor of Physics Emeritus at the California Institute of Technology in the USA. He was first the Principal Investigator of the LIGO project and later the Director. He led the efforts in the final design, approval for funds and the construction and commissioning of the LIGO interferometers at Hanford and Livingston. He also set up the LIGO Science Collaboration (see box). Kip S. Thorne is the Feynman Professor for Theoretical Physics Emeritus at the California Institute of Technology. He is known for his work on gravitational physics and astrophysics. He developed the formalism for the analysis of the generation of gravitational waves, contributed to the development of plans for gravitational wave detection and is a co-founder of the LIGO project. He is a co-author with John Wheeler and Charles Misner of the book "Gravitation" which has been very influential. He was an executive producer and scientific adviser for the science fiction movie "Interstellar" by Christopher Nolan.

LIGO Science Collaboration
Setting up the LIGO detectors was an extremely difficult task which took many years to accomplish. The task required research and development in many areas of astronomy, physics and engineering. The analysis of the data obtained again requires enormous efforts. All this work cannot be done by a small number of people. It is necessary to have a large number of research groups, comprising of hundreds of individual scientists, working together over a long period of time. To provide a platform for such large and sustained collaborations, the LIGO Science Collaboration (LSC) was set up in 1997 under the leadership of Professor Barry Barish of the California Institute of Technology. Its mission is to detect gravitational waves, use gravitational waves to explore fundamental physics of gravity and to develop gravitational wave observations as a tool of astronomical discovery. Individuals as well as groups and institutions anywhere in the world who can contribute to the mission of the LSC can become its members. In 2019, the LSC had over 1200 members from over 100 institutions in 18 countries. LSC members who spend a significant fraction of their time on research in gravitation waves become authors of papers published under the LSC. The LSC also has access to the data from the Advanced VIRGO gravitational wave detector in Italy and LSC members work closely with the VIRGO Collaboration.

8.3   Other Gravitational Wave Detections

The first observing run of the aLIGO detectors, designated O1, lasted from 12 September 2015 to 19 January 2016. During this run a total of three black hole binary merger events were detected. After detector and system upgradation, the second observing run O2 began on 30 September 2016 and continued until 25 August 2017. During this period eight detections were made, including a binary neutron star merger, which was also detected at electromagnetic wavelengths. This will be described in detail in Sect. 8.5. We will briefly describe below four of the detected sources, GW151226, GW170104, GW170608 and GW170814. Of these, the first was detected in O1, while the next three were detected in O2. The source GW170729 detected in O2 is the most distant and luminous source observed so far. The merger occurred about 5 billion years ago, that is before the formation of the Solar system, and about 5 Solar masses were lost from the system in the form of gravitational waves emitted during the merger. The merger GW170814, which we describe below, was the first source to be observed by a three detector network (aLIGO Livingston and Hanford and Advanced VIRGO), which allowed for the first tests of a property of gravitational waves known as polarisation.

In addition to the confirmed detections, there have been several events which led to signals which could have been due to gravitational waves. But in these cases, the signals were not strong enough for them to be accepted with sufficient confidence as being real sources and not spurious signals due to noise in the detectors.

GW151226: This is again a black hole binary source which merged to form a single black hole. It was detected, as its name implies, on 26 December 2015, i.e. on Boxing Day, which is the day following Christmas Day. It was observed at 03:38:53 UTC by the LIGO detectors at Livingston and Hanford, the inferred merger time at Livingston being 1 ms earlier than at Hanford. The signal in this case was weaker than in the case of GW150914, and the event was spread over 1 s compared to the 0.2 s spread of the earlier detection. As a result, it was more difficult to detect the signal, and a special technique known as matched filtering had to be used. This was first developed in the context of gravitational wave detection by Bangalore Sathyaprakash and Sanjeev Dhurandhar. As can be seen from Fig. 8.4, over the 1 s that it was detected, the signal went through 55 cycles, during which the height of the peak increases with time, as does the frequency, since the two components of the binary spiral towards each other due to the energy loss to gravitational wave emission. The peak frequency reached is 450 Hz, which means that the two components approach each other very closely and must therefore be black holes. Calculations show that the masses of the black holes are about 14.2 and 7.5 Solar masses, which makes them less massive than the black holes in GW150914. The mass of the spinning black hole produced through the merger is about 20.8 Solar masses, so that about 1 Solar mass is lost from the system due to the emission of gravitational waves. The source is at a distance of about 1.4 billion light years from us, which means that the merger detected by LIGO on 26 December 2015 occurred 1.4 billion years ago! The modelled waveform of

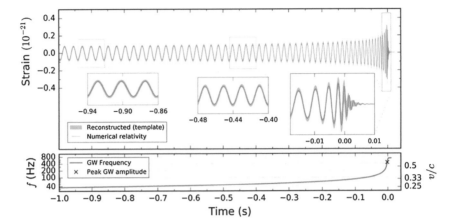

Fig. 8.4 GW151226. Upper panel: The waveform as estimated from theory is shown, together with numerically calculated models of the black holes at various stages of the eventual merger. Lower panel: the increase in frequency as the components spiral-in, and the consequent increase in their relative velocity are shown. Image credit: B. P. Abbott et al. PRL **116** 241103, 2016

GW151226 is shown in Fig. 8.4. The ringdown to single black hole formation is clearly seen in the inset on extreme right.

GW170104: This source was detected on 4 January 2017, and was the first source to be found after the start of the second observing run of Advanced LIGO in November 2016. Detailed analysis as in the earlier cases showed that it was a black hole binary which merged to form a single black hole, with masses of the two components in the initial binary system being 31.2 and 19.4 Solar masses. The final black hole has mass 48.7 Solar masses, so that about 2 Solar masses is lost from the system upon merger due to the emission of gravitational waves. The source is at a distance of about 2.9 billion light years.

GW170608: This source was detected on 8 June 2017. It was again a black hole binary which merged to form a single black hole. Detailed analysis of the data showed that the masses of the two components in the binary system were 12 and 7 Solar masses, respectively, while the mass of the black hole formed from the merger is 18 Solar masses. The difference of 1 Solar mass was radiated away in the form of gravitational waves. The source is at a distance of 1.1 billion light years. The masses of the black holes in this source are considerably smaller than the masses in other sources observed so far.

GW170814: This is a black hole binary detected on 14 August 2017. The calculated masses of the black holes in the binary are about 30.5 and 25.3 Solar masses, while the mass of the black hole after the merger is about 53.2 Solar masses. The mass lost from the system due to the emission of gravitational waves is about 2.6 Solar masses. The source is at a distance of about 1.8 billion years from us. The novelty in the detection is that for the first time a source was observed from three locations: the

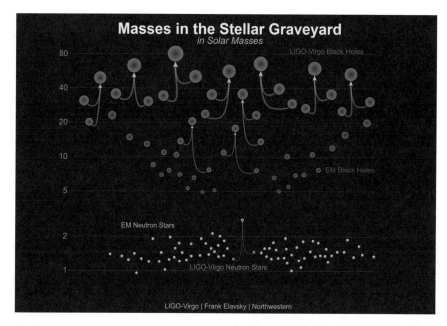

Fig. 8.5 The 10 black hole binaries and one neutron star binary whose mergers were observed in observing runs O1 and O2. Details of the figure are explained in the text. Image credit: LIGO/Frank Elavsky/ Northwestern

two Advanced LIGO detectors at Livingston and Hanford and the Advanced VIRGO detector in Italy. This enabled the direction of the source in the sky to be limited to a region of 60 square degrees. This is much smaller than the possible regions in the sky for the other three sources, which were observed only by the two Advanced LIGO detectors. Using observations from the three detectors it has been possible to measure gravitational wave polarisation. These measurements are fully consistent with the prediction of general relativity, and are not consistent with results expected from other competing theories of gravity.

The 10 black hole binaries and one neutron star binary whose merger was detected in the observing runs O1 and O2 are shown schematically in Fig. 8.5. For each source, the blue circles, labelled LIGO-Virgo black holes in the figure, indicate the mass of each component of the binary before merger and the mass of the remnant post-merger. The purple circles indicate the mass of black holes detected as a component of an X-ray binary system (see Sect. 5.4.3.1). These black holes are detected through X-ray and optical observations of the binary systems, i.e. through electromagnetic waves, so the purple circles have been labelled as *EM black holes*. The yellow circles indicate the masses of neutron stars detected electromagnetically *(EM neutron stars)* in X-ray binary systems in which the compact component is a neutron star.

It may be noticed from the diagram that the masses of the LIGO-Virgo black holes in many cases are significantly larger than the masses of the EM black holes. The reason is that the larger the black hole mass, the greater is the gravitational wave luminosity of the system and shorter is the time over which the signal is spread out at the merger. This makes the signal more easily detectable over the background noise. It is expected that there will in fact be many more low mass black hole binaries than the high mass binaries which have been detected so far, and the lower mass systems will be detected with increased sensitivity of the detectors.

The EM neutron star masses in Fig. 8.5 are all well below the Tolman-Oppenheimer-Volkoff maximum mass limit for neutron stars, which is about three Solar masses as discussed in Sect. 5.4.2.1. In the case of the binary neutron star merger event GW170817, the mass of the remnant is estimated to be 2.82 Solar masses (see Sect. 8.5), which is at the upper end of the permissible mass of neutron stars. The remnant could very well be a black hole. In that case, it would be the only black hole which is in the mass range of 2–5 Solar masses. This region is known as the *mass gap* region because of the lack of black holes in it. Binary black holes with pre- or post-merger component mass in this region would be difficult to detect through their gravitational wave emission, given the present sensitivity of the detectors, and none has been found through electromagnetic observations. It is possible of course that there are no black holes, or at least relatively few black holes, with mass in this region. That would be a very interesting astrophysical fact which would need to be carefully examined and explained.

8.4 Binary Black Hole Formation

The five gravitational wave sources which were announced until October 2017 were all binary black holes. Such binaries, with massive black hole components which orbit each other at close distance are the easiest to detect, because they are copious emitters of gravitational waves at frequencies which are detectable by aLIGO. They were not seen by the initial LIGO detectors. But the increased sensitivity of aLIGO, which is at least 3–4 times higher than the sensitivity of initial LIGO in the range of frequencies detected, allowed the binaries to be observed in the last stages of their evolution, as they spiralled towards merger to a single black hole. Only the last stage, which lasted for a fraction of a second, was seen as only at this stage the emission of gravitational waves was powerful enough to be detected over distances greater than a billion light years.

The detections establish that there is a considerable population of such black hole binaries in the Universe. The number of detections so far is too few to make an accurate estimate of how many mergers we can expect to occur per year over even the distances measured for the sources and over greater distances. But under various assumptions, rough estimates can be made as a start. The result obtained from such calculations is as follows: consider a large cube with each side being 3.3 billion light years in length. We can expect between 2 and 50 mergers to happen in such a volume per year. Apart from the detections, there have been several signals observed which could be due to gravitational wave sources, but which did not have enough signal strength to be confidently accepted as being real sources rather than artefacts. If these were indeed real sources, then the number of mergers in the large cube would go up to a few hundred per year. Either way, we can expect that aLIGO will detect a steady stream of sources providing us with completely new insights.

Binary black holes had not been observed before the LIGO detections. But binaries in which one of the components is inferred to be a black hole have been known from observations of X-ray binaries, as discussed in Sect. 5.4.3.1. In these sources, one component is a more or less normal star from which matter flows onto the other component, which is a neutron star or a black hole. During the flow the matter is heated to high temperature and emits X-rays through a variety of processes. The mass of the compact object can be estimated, and if this turns out to be greater than about 2–3 Solar masses, it must be a black hole, as explained in Sect. 5.4.2.1. More than 20 X-ray binaries with the compact object a black hole are known in our galaxy. A majority of these black holes have mass in the range 5–10 Solar masses, while some others have higher masses in the range 10–20 Solar masses.

In the five binary black holes detected so far, the masses of the ten black holes are 36, 29, 14.2, 7.5, 31.2, 19.4, 12, 7, 30 and 25 Solar masses. At least five of these black holes are significantly more massive than the range of black hole masses in X-ray binaries, and only two black holes are well within the range. So the very first gravitational wave detections have led us to unexpectedly massive black holes, with the added surprise that they reside in binary systems. Astrophysicists have still to fully understand the ways in which such systems could be formed.

As discussed in Sect. 5.4, when the nuclear fuel in a star is completely exhausted, the central region of the star collapses, while the outer region expands and is thrown out in an explosion. The ultimate fate of the collapsing core depends on its mass: when the core mass is less than the Chandrasekhar limit of 1.4 Solar masses, it forms a white dwarf. When the core mass is greater than the Chandrasekhar limit, but less than the Tolman-Oppenheimer-Volkoff limit of about three Solar masses, a neutron star is formed, and when the core mass is greater than this limit, we get a black hole. The mass of the collapsing core increases with the initial mass of the star, at least for the collapses that astronomers have been familiar with. The formation of a black hole requires the star to be more massive than about 25 Solar masses.

The massive black holes found in the gravitational wave emitting binaries require very massive stars indeed, whose evolution from initial stages to collapse is not well understood at the present. In particular, the familiar relation between initial star mass and core mass may break down. It is possible for some of the exploding portion to be dragged back to collapse with the core. It is also possible that the whole star implodes to a massive black hole. The details depend on how much mass the evolving star loses due to winds blowing from the surface, which in turn depends on the fraction of heavy elements in the gas comprising the star. These are all unknown for the stars required to form massive black holes, and much research is in progress on these matters.

The next question is how does one form binary black holes? We could begin with a binary of two massive stars each of which evolves to the final formation of a massive black hole. As in the case of the formation of a neutron star binary we considered in Sect. 6.3, the formation of the black hole binary would require considerable exchange of matter between the stars, and loss of matter from the system. If these did not happen in the correct way, the initial binary would be disrupted when either the first or the second explosion takes place during the formation of the black hole. When the black hole binary is formed, it is required that it is sufficiently compact. The rate at which gravitational waves are emitted depends on the distance between the two black holes. With the emission of gravitational waves the binary shrinks in size, and eventually the two components merge together producing an event like GW150914. If the black hole binary is too wide when it is formed, the emission of gravitational waves will be negligible and the time taken by the binary to merge will be even longer than the present age of the Universe. We will of course never see such binaries merging!

Binary black holes could also be formed through another route, in which two black holes formed separately capture each other to form a binary. This can happen only in a dense environment where there are many black holes present, so that they can approach each other closely enough for capture. Such dense environments are found in the central regions of galaxies and globular clusters, which are large aggregates of stars with high density of stars in their central regions. The evolution of massive stars in such regions would lead to the formation of black holes which could then capture each other. When the binary is formed it has negative energy, while the two black holes initially have positive energy. So the formation requires some third body to carry away the energy. While such processes are possible in principle, the details are still being investigated.

8.5 Binary Neutron Star Detection

On 17 August 2017, the Advanced LIGO detectors at Hanford and Livingston, and the Advanced VIRGO detector in Italy, detected a signal which lasted about 100 s, which has been named GW170817. From simple visual inspection of the two Advanced LIGO signals, it was apparent that this signal could have been generated by the spiral-in of a neutron star binary. Moreover, a cosmic Gamma-ray burst (see below) was independently observed by the Gamma-ray burst Monitor (GBM) on the Fermi satellite just 1.7 s after the end of the gravitational wave signal. A Gamma-ray burst is indeed expected when the neutron stars in a binary merge together at the end of the spiral-in to form a single object. A worldwide alert was therefore generated so that astronomers could prepare to observe the event with ground-based and space telescopes. Quick analysis of data from the three gravitational wave detectors made it possible to locate the source within an area of about 31 square degrees.

In the case of the black hole binaries detected by aLIGO, the emitted gravitational waves could be observed only for a fraction of a second before the merger of the black holes. Only in this stage the power emitted by the system was sufficiently large and had the correct frequency range to be detected by Advanced LIGO. After the merger and formation of the single black hole, a ringdown phase was seen as the black hole settled down to its final state. Neither the component black holes before the merger, nor the final black hole emit any electromagnetic radiation. The process therefore cannot be observed by any telescope for electromagnetic waves and all our knowledge of the black hole systems comes from the observation of gravitational waves alone.

Neutron star binaries are quite different from black hole binaries. In Chap. 6, we have discussed in detail the binary pulsar PSR1913+16, in which one component of the binary is a neutron star which is radio pulsar, while the other component is a non-pulsating neutron star. It is the radio pulsations which enabled the discovery of the object, and accurate measurements of the small changes in the observed period led to the determination of the properties of the neutron stars and the binary. It was observed that the size of the binary is shrinking, which could be attributed to energy loss due to the emission of gravitational waves by the binary. As the binary becomes more compact, the rate at which energy is lost increases, so that the binary shrinks in size even faster. It is estimated that the spiral-in process will be completed and the neutron stars will merge in about 300 million years.

The process of merger of the two neutron stars is very violent. As the stars approach each other ever more closely, the gravitational field of each neutron star distorts the shape of the other star, like the tides produced in the oceans of the Earth due to the gravitational fields of the Sun and the Moon. As the distance between the two neutron stars becomes comparable to their radius, which is about 10 km, the tidal force becomes so large that each star is torn apart and the matter from the two merges together. The gravitational energy released in the process leads to a tremendous explosion in which some of the matter is ejected from the system. The remaining matter undergoes a collapse due to the large gravitational force dragging the matter inwards. If the mass of the collapsing core is less than the maximum permissible mass of a neutron star, then as discussed in Sect. 5.4.2.1 the collapsing object would form a neutron star. But if the mass of the remnant is more than the maximum mass limit for a neutron star, a black hole must be formed from the merger.

The explosion which results during the neutron star merger is expected to produce a Gamma-ray burst. In such bursts a tremendous amount of energy is released, in which first there is explosive release of Gamma rays, which is followed by electromagnetic radiation of various kinds, including X-rays, optical radiation and radio waves. Hundreds of Gamma-ray bursts lasting for various duration, ranging from a fraction of a second to hundreds of seconds and longer have so far been observed. The amount of energy emitted in such a short time is equivalent to the total energy emitted by a star like the Sun in about a trillion years. Some of the bursts with short durations of less than 2 s are believed to be produced by merging neutron stars.

The detection of the Gravitational wave source GW170817 and the independent detection of a Gamma-ray burst within 1.7 s of it generated tremendous excitement, as the event could lead to better understanding of gravitational wave sources, Gamma-ray bursts and very high-density matter in its most extreme form. On the gravitational side, analysis of the data from the Advanced LIGO and Advanced VIRGO detectors led to very interesting results. The total mass of the system before the merger was found to be in the range of 2.73 and 2.82 Solar masses, the mass of one of the components was in the range of 0.86 to 1.36 Solar masses, while the mass of the other component was in the range of 1.36 to 2.26 Solar masses. The masses cannot be more accurately determined because these are linked to how much spin the two objects have, and that is not known at the present. Since the two components have merged and do not any longer exist as independent objects, their further observation is not possible. But over time it will be possible to analyse the data in increasingly sophisticated ways, so that more information will be obtained and used to determine the properties more accurately.

A very important quantity associated with the source is its distance from us, and that is determined from the gravitational wave observations to be about 130 million light years, which is much less than the distance to the four detections described above. As a consequence GW170817 is the most intense gravitational wave source detected so far. Because it was so intense, the source could be detected by the Advanced VIRGO detector as well, in addition to the two aLIGO detectors, which allowed the position in the sky to be determined to within an area of 31 square degrees, as mentioned above. The source is near the southern end of the constellation Hydra. Astronomers searched this part of the sky with a multitude of telescopes and soon found a transient source which had appeared in the galaxy NGC3993, about 75,000 light years from its centre. Detailed observations of the source were carried out at optical, near-infrared, X-ray and radio wavelengths by various ground- and space-based telescopes and a great deal of data was obtained.

What was the nature of the of the two components of the binary before the merger? The mass of the two components was well within the range of known neutron star masses which were determined from the many known binary neutron stars and X-ray binaries. The two masses were also well below the mass of known black holes associated with binary stellar systems, as determined from gravitational wave and X-ray data. It is therefore reasonable to assume that the two components were neutron stars, and much of the work on the source has proceeded on that assumption. Nevertheless, it is important to keep in mind that either of the objects could have been compact exotic objects like quark stars or they could be low mass black holes. Further information will be obtained as the analysis of data on GW170817 progresses and more such systems are observed in the future. But it is clear that both objects could not have been black holes. The observation of Gamma rays and other electromagnetic radiation from GW170817 means that at least one of the object must have had finite size with matter flow having taken place from this object to the other one. If both components were black holes then no electromagnetic radiation would have been detected.

What is the nature of the remnant formed from the merger? The mass of the remnant, which is 2.82 Solar masses, is again in the range of known neutron star masses, but close to the upper limit. So while a large mass neutron star could be formed, it may be short lived, collapsing in less than a second, because the mass cannot be supported any longer by the pressure of the neutrons. The neutron star could also be long lived, existing for 10,000 s or even much longer duration, before collapsing to a black hole. A neutron star remnant would emit gravitational waves upon formation because of irregularities present in its structure, but these are at high frequencies which cannot be efficiently detected by aLIGO. The merger could also lead to the formation of a black hole, in which case there would be the ringdown gravitational wave emission as in the case of GW150914. But this again is at high frequencies around 6000 Hz, at which the detectors do not have sufficient sensitivity. Searches have been made in the data for gravitational wave emission at somewhat lower frequencies, but no conclusions can be drawn as of the end of 2019 regarding the nature of the remnant.

Observations of the electromagnetic counterpart of GW170817, known as EM170817, at various wavelengths have led to a great deal of information about the nature of the spectrum in the optical and near-infrared regions. From the shape of these spectra, it can be concluded that several elements heavier than iron must be present in the matter expelled during the merger. The production of such elements has so far remained a puzzle. It is known that the lightest elements hydrogen and helium and some light elements like lithium, beryllium and boron in trace quantities are produced in the Big Bang. The elements heavier than helium continuing up to iron are produced in the interiors of stars due to nucleosynthesis. It has long been believed that elements heavier than iron, particularly those which require a mechanism involving the rapid capture of neutrons by atomic nuclei, were produced in supernovae. More recently it has been argued that such elements could be produced in the neutron-rich matter present during neutron star mergers. The observed spectra of EM170817 are consistent with these expectations, which is an important development. Some of the nuclei produced in this manner are unstable and undergo radioactive decay, leading to explosive electromagnetic emission which is known as a kilonova. The observed emission from EM170817 is consistent with known theoretical models of kilonovae.

A number of short-lived Gamma-ray bursts have so far been observed. It is believed that the emission from such sources is produced by matter moving in a narrow jet at speeds very close to the speed of light towards the observer. It has also been believed that such short period Gamma-ray bursts are produced during the merger of neutron stars. Observation of the emission from EM170817 shows that while it is indeed a short duration Gamma-ray burst, the rate of Gamma ray emission from it is about 10,000 times smaller than other such known bursts. Detailed considerations show that EM170817 must be quite different from the earlier sources, and new models are required to explain its behaviour. While several models have been considered, there is no agreement yet as to the correct one which is consistent will all observations.

Observations of the neutron star binary lead to other results of importance to cosmology and fundamental physics. The distance to the binary is determined from the gravitational wave observations, as discussed in Sect. 8.2.1, while the cosmological redshift is determined from the optical spectrum of the electromagnetic counterpart. Combining the two leads to determination of Hubble's constant, as explained in the box on cosmological redshift in Sect. 5.4.3.1. Hubble's constant obtained in this manner is consistent with the value obtained by conventional means which use only electromagnetic observations. With further neutron star binary observations, the precision with which Hubble's constant is determined from gravitational wave observations will improve. The Gamma-ray burst associated with the neutron star merger was observed 1.74 s after the merger. From the delay, it is possible to determine that the speed of gravitational waves is the same as the speed of light, to within one millionth of a billionth of the light speed, confirming with great precision Einstein's prediction that the two speeds must be the same.

The observation of the gravitational radiation from the GW170817 and the associated electromagnetic radiation have been very exciting developments. It was believed that the first detection by aLIGO would be of merging neutron stars. But the actual first detection was of a binary black hole of unexpectedly high component masses. The wave pattern beautifully matched theoretical expectations for such an object, down to the elegant ringdown pattern. While the gravitational wave pattern from GW170817 is consistent with the expectations from the spiral-in and merger of a neutron star binary, there could still be some surprises in store. The associated electromagnetic emission clearly calls for new theoretical studies, while it seems to be capable of explaining some long-standing issues like the formation of some elements heavier than iron. The gravitational wave detectors are proving to be great sources of new physics and astrophysics.

8.6   Observing Run O3

The third observing run O3 of the Advanced LIGO and Advanced VIRGO detectors started on 1 April 2019. Beginning with this run, information about detected signals which are thought to be of astrophysical origin is quickly made public, so that all interested scientists and other persons can have access to the information. This is done through the *Gamma-ray Coordinates Network (GCN*, https://gcn.gsfc.nasa.gov), which is a portal for discoveries and observations of astronomical transients, including those at electromagnetic wavelengths and gravitational wave, cosmic-ray and neutrino events.

A number of software pipelines analyse the data from the two aLIGO detectors and the Advanced VIRGO detector in real time as it is collected, searching for gravitational wave signals in the data. When a signal which passes certain criteria is detected by the pipelines, the information is automatically sent to the scientific community in the form of a *Preliminary GCN Notice*. The quick notice is issued so that electromagnetic observations by interested groups can begin immediately. Following that, human experts take a more careful look at the data and if the signal successfully passes scrutiny, an *Initial GCN Alert* is issued, followed by a *GCN Circular*, which is taken to be the first formal publication of a gravitational wave candidate event. The Notices are machine readable, while Circulars are human readable astronomical bulletins. At this stage, the estimated source location in the sky, the probability that the event is one of several types like a black hole or neutron star merger, the false alarm rate, which is the number of years in which one such event could falsely be detected, and some limits on the mass are released to the public. If the signal does not pass human scrutiny, then a *Retraction GCN Notice* is issued as the signal is no longer considered to be of astrophysical origin and is very likely to have been due to noise in the system.

After the acceptance of a signal as a candidate gravitational wave source, it can take several months or even longer for the signal to be accepted as a confirmed gravitational wave source. This is because of the multiple tests that the signal is subjected to, and the calculations required for a careful estimation of the parameter values. A signal is taken to be finally confirmed when a research paper on the finding is accepted for publication by an international research journal, after careful consideration by one or more independent referees.

Information about candidate events is made publicly available on the Gravitational Wave Candidate Event Database (GraceDB, http://gracedb.ligo.org). As of the middle of February 2020, GraceDB has 52 candidate events. A number of other sources which were considered to be candidates have been retracted after further study. Of the candidate sources, (1) 20 events have been tentatively classified as binary black hole (BBH) mergers with probability exceeding 99% and another 12 with probabilities exceeding 92–98%; (2) three events have been tentatively classified as mergers of binaries consisting of a black hole and a neutron star (NSBH), with probability exceeding 93, 98 and 99%, respectively; (3) one event has been tentatively classified as the merger of a binary neutron star (BNS) with probability greater than 99% and (4) two events have been tentatively identified as being mass gap sources (MassGap) with probability exceeding 99% and another as a mass gap source with probability exceeding 95%. Some of the remaining sources have been tentatively identified with binaries of one of the four kinds, but with smaller probabilities, while the rest are classified as sources of terrestrial origin with very high probability. None of the candidates, including the binary neutron star, have been identified with an electromagnetic counterpart. The expected nature of O3 candidate sources is summarized in Fig. 8.6.

During further work on confirming the identifications as of April 2020, a very interesting discovery has been made. The source GW190412 has been confirmed to have been a merging black hole binary system, with rather unequal black hole masses. Prior to the merger, the masses of the black holes were 30 and 8 Solar masses, so that ratio of the two masses was 3.6. The component masses in the other gravitational wave detections of merging binaries identified so far have not been so unequal. The ratio of the larger mass to the smaller mass in the other known binary black hole mergers is less than 1.8.

Component mass differences have an effect on the waveform emitted by the binary and these differences increase as the masses become more unequal. We have mentioned earlier that the frequency of the gravitational waves emitted by a binary is twice the frequency of rotation of the components. This is true when the two components have equal mass. When the masses become unequal, there is also emission at higher frequencies, known as the harmonics of this fundamental frequency of emission. The energy in the harmonics increases as the masses become increasingly unequal. The mass difference in GW190412 was large enough for higher harmonics to be detected. The energy in the first harmonic, which has a frequency three times the orbital frequency, is consistent with the prediction of general relativity. The higher harmonics have a practical application: they can be used to better constrain the distance to the

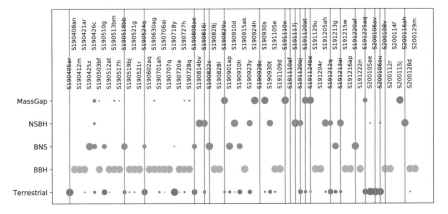

Fig. 8.6 Possible nature of some of the candidate gravitational detections from O3. The names of the events are shown at the top of the figure. The circle below each name indicates the possible type of the event, with different colours indicating black hole binary (BBH), neutron star-black hole binary (NSBH), neutron star binary (BNS), mass gap binary (MassGap) or terrestrial events. The size of the circle is indicative of how high the probability is of the event being of a particular type. More than one circle is present under some events, when the probability is significant for more than one type. Image credit: Varun Bhalerao

binary system and how much the plane of rotation of the two components is inclined to the line of sight.

The observing run O3 was ended on 27 March 2020, about a month before the planned shutdown, because of the situation created by COVID-19. The next run O4 is expected to begin in early 2022 and will last for a year. In the period between O3 and O4, various systems in the interferometer will be upgraded, so that the design sensitivity of aLIGO will be reached in O4. The distance to the neuron star binary GW170817 was 130 light years. With the sensitivity attained during O3, such a binary could have been observed to a distance of about 400 million light years. With the improved sensitivity during O4, this distance increases to about 600 million light years. After the end of O4 there will be a period of about two years during which the interferometer will be upgraded to A+, as described in Sect. 7.5.5. It is expected that during O5, which may begin sometime in 2025, a GW170817 like binary will be observable to a distance of 1 billion light years. The volume of space over which such a source will be observable during O5 will be about 15 times the volume which was covered during O3. The number of observable sources of all kinds will therefore be much larger, and more precise data will be available. Advanced VIRGO will also be upgraded on a similar schedule. O5 will be followed by LIGO Voyager.

Chapter 9
Future Gravitational Wave Detectors

Abstract In this chapter, we discuss the future of gravitational wave detectors beyond the present second generation (2G). Several ambitious third generation (3G) projects which will significantly increase the reach of the detectors, through increased sensitivity as well as the ability to detect gravitational waves with lower frequencies, are being planned. These include the Cosmic Explorer, which will be like the LIGO detector but with 40 km long arms and the Einstein telescope, which will have three 10 km long arms in a triangular configuration. The minimum frequency to which these detectors will be able to observe is limited to about 5 Hz. Space-borne observatories are needed to probe much lower frequencies to detect sources like massive black hole binaries and a gravitational wave background. We describe three such missions, LISA, TianQin and DECIGO, as well as the LISA Pathfinder which was launched to test technologies needed for LISA. We end by describing Pulsar Timing Arrays for the detection of gravitational waves with frequencies as low as a hundred millionth of a Hertz.

9.1 Introduction

Following the first detection of gravitational waves from a black hole binary in 2015, more such binaries have been detected in the first two observing runs of aLIGO and Advanced VIRGO. The merger of a neutron star binary was also detected, accompanied by the observations of the electromagnetic counterpart. All these observations have led to highly interesting scientific results. In observing run O3 by the gravitational wave detectors, candidate detections were made about once a week, and further study of these sources will lead to many more discoveries. A new field of gravitational wave astronomy has therefore been established, and yet the observations made so far have been limited to the mergers of compact binary sources. These sources are the easiest to detect with the available facilities. Going beyond them will require the development of more sensitive detectors and the exploration of lower frequency gravitational waves.

Greater sensitivity will enable the detection of fainter sources, which would include bright sources which are at great distances, up to the early epochs of the

Universe. That would help us study how the nature and number of sources, for example, black hole binaries, changed as the Universe evolved. Sources like core-collapse supernovae, single neutron stars, white dwarf binaries or black hole binaries in the early stages of their evolution when their orbital radius is large, are much fainter then the sources detected so far. With increased sensitivity, it will be possible to detect such sources, which will be abundant in the galaxy. We have seen in Sect. 8.2.1 that the waveform of the first binary black hole merger detected, GW140915, was broadly in conformity with the predictions of general relativity, right up to the final ringdown phase after the merger. Greater sensitivity will enable such waveforms of gravitational waves to be studied ever more accurately to test in detail various predictions of general relativity about black hole physics. Any departures found will be of fundamental importance, since they could indicate the necessity of a new theory for very strong fields, the presence of exotic objects or they may even be signatures of the quantum nature of gravity.

The development of detectors which are sensitive to gravitational waves of lower frequencies than presently observed will again be very exciting, as contributions to these bands come from a variety of sources like supermassive black hole binaries, wide stellar mass black hole binaries, the stochastic gravitational wave background, the inflationary epoch of the very early Universe and so forth.

The sensitivity of aLIGO like detectors can be improved by increasing the arm length of the interferometer as mentioned in Sect. 7.5.5. The sensitivity can also be improved by reducing different kinds of noise in the system and improving the optics. There are plans to build 3G (third Generation) detectors with arm lengths as long as 40 km, placing detectors under the ground to reduce seismic and Newtonian noise, to use mirrors cooled to low temperatures to reduce thermal noise and to use higher powered lasers to reduce the quantum noise. Plans have also been made to have detectors in a three arm triangular configuration for better sky coverage and simultaneous detection of both polarisations of the gravitational waves. These projects aim to provide about ten times better sensitivity than the current 2G (second Generation) detectors, and the ability to detect gravitational waves to 5 Hz.

Increasing the arm length even further and probing lower frequencies will not be possible on the Earth. For achieving these improvements, space-borne detectors will be needed. In space, the arm length can be increased almost indefinitely by using multiple satellites. The thermal environment in space is more stable than on the ground, and the seismic and Newtonian noise, which affect detector performance at low frequencies, are completely absent. These factors will allow frequencies as low as 0.1 mHz to be detected. Probing even lower frequencies, around 10^{-8} Hz, is possible by observing the effect of gravitational waves on the arrival time of pulses from millisecond radio pulsars.

In the following, we will describe some of the detectors planned for the future. We will first describe two terrestrial projects, *Cosmic Explorer* and the *Einstein Telescope*. Then we will describe the space-missions *LISA*, *LISA Pathfinder*, *TianQin* and *DECIGO*, and end with *Pulsar Timing Array*. The gravitational wave frequency regions and sources explored by existing detectors and these missions are shown in Fig. 9.1.

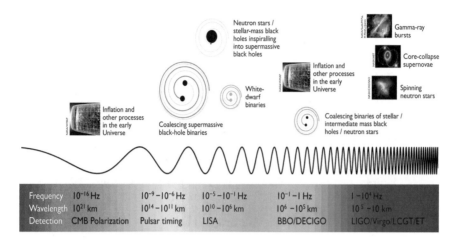

Fig. 9.1 The gravitational wave frequencies emitted by various sources, and the detectors which can be used at those frequencies are schematically indicated. The Cosmic Explorer will occupy a niche similar to the Einstein Telescope. LCGT is the Large-scale Cryogenic Gravitational wave Detector, now renamed as KAGRA. BBO, the Big Bang Observer, is a possible successor to LISA. Image credit: Ajith Parameswaran

9.2 Einstein Telescope

The Einstein telescope (ET) is a proposed 3G gravitational wave detector to be located in Europe. A design study of the telescope has been completed and a detailed proposal is in preparation as of 2020. It is expected that the telescope will be ready sometime in the 2030s.

The Einstein Telescope will be triangular shaped, with three arms. Each arm will be 10 km long, making it two and a half times as long as an arm of the LIGO detector. The whole telescope will be buried in the ground, a few hundred meters below the surface. The increased length of the arms helps in making the telescope more sensitive by reducing the effect of noise which causes displacement of the mirrors.

The three-arm design will make the telescope more equally sensitive to sources distributed in the sky, than a telescope with two arms. The three arms will also help in the measurement of the polarisation of detected gravitational waves. The telescope will be built underground because that reduces seismic noise and thermal noise, which we have described in Sect. 7.5. These sources of noise are dominant below a gravitational frequency of about 100 Hz, so their reduction will make the telescope more sensitive than aLIGO at lower frequencies. An artistic view of ET is shown in Fig. 9.2.

We have seen in Sect. 7.5 that quantum noise dominates above 100–200 Hz. The reduction of this noise requires the use of high-powered lasers. This increases the intensity of radiation in the arms, providing a larger number of photons and therefore less quantum noise. But the high power can cause heating of the surface

Fig. 9.2 An artistic view of the Einstein telescope located underground. Image credit: ET (Einstein Telescope) Project

of the mirrors, leading to greater thermal noise in the mirrors and their suspensions. The heat can be removed and the thermal noise reduced by using a cooling system to keep the mirrors at a temperature of about 20 K. But that requires the thickness of the suspension wires to be increased to help remove the heat, which reduces the sensitivity of the detector at lower frequencies.

To maintain optimal sensitivity at low as well as high frequencies, ET will have two separate detector systems nested in the same enclosure. One system will have relatively low laser power and a cooling system for the mirrors. This will have high sensitivity at lower frequencies in the range of 2–40 Hz. The other system will use high laser power to reduce the quantum noise, and will be maximally sensitive at the higher frequencies.

Due to its increased arm lengths, three arms, the blending of two detector systems and improved technology, ET will have significantly greater sensitivity than aLIGO at its best. ET will also be able to observe over a larger range of gravitational wave frequencies. In terms of the cosmological redshift (see the Cosmological Redshift Box in Sect. 5.4.3.1), ET will be able to observe binary black holes, of the kind observable by aLIGO to a redshift of about 20. For aLIGO, the maximum redshift for such binaries is limited to about 1.0. The larger redshift range accessible to ET means that for the black hole binaries it will be able to observe over a volume about 35 times larger than the volume accessible to aLIGO, thus increasing the number of sources which can be observed. Since the farther a source is, the greater is the time taken by light to reach us from the source, the farthest binaries which could be observed will have emitted their signal about only 100 million years after the Big Bang. Any such black hole binaries detected will be of primordial origin, i.e. they would have formed before the first stars formed. ET will therefore be able to detect binary black holes over the entire history of the Universe, from the present to the earliest times. Because it can detect gravitational waves at lower frequencies than LIGO, ET will be able to detect binaries with black hole masses reaching about thousand Solar masses. ET will also be able to detect other gravitational wave sources like core-collapse supernovae, neutron stars with deformities and stochastic sources which we have described in Sect. 5.5, and various objects and phenomena associated with the early Universe which so far remain inaccessible.

9.3 Cosmic Explorer

This is a proposed 3G gravitational wave detector to be based in the U.S. It will be developed through two stages, Cosmic Explorer 1 (CE1) to be ready in the 2030s and Cosmic Explorer 2 (CE2) to be ready in the 2040s. A technology development programme needed for the project has been proposed. CE1 will have two arms in an L-shaped configuration like LIGO, but in this case each arm will be 40 km in length, with a tube diametre of about 1 m. The ten-fold increase in arm length is the main contributor to the order of magnitude increase in sensitivity expected of the detector. But then a large suitable site will have to be carefully identified to accommodate the very long arms. Over a 40 km distance the curvature of the Earth becomes significant and levelling to an extent of 30 m will be required, depending on the details of the site. The beam tubes would run on the surface, below the surface or above the surface at different locations of the site. Creating the ultra-high vacuum over the large volume of the tubes at reasonable cost, and to make it last for some decades will require innovative technology.

CE1 will use the technology developed for the LIGO A+ detector, with enhancements appropriate for the larger beam tube length and higher sensitivity. For example, the CE1 test mass mirrors will weigh 320 kg while the A+ test mass will be 40 kg. In both detectors, the mirrors are made of fused silica. The circulating laser power in the CE1 beam tube will be 1.4 MW while in A+ it will be 0.8 MW. CE1 will be sensitive over a frequency range of 5–4000 Hz, making it possible to detect lower frequency sources not accessible to LIGO A+. CE1 will be able to detect binary neutron stars, of the type detected by aLIGO, to a redshift of about 3, compared to the redshift of about 0.17 that LIGO A+ will be able to reach. Due to its increased sensitivity, CE1 will be able to provide more than 10 times stronger signals compared to the noise than LIGO A+, so that the shape of the signal as it changes with time can be more accurately measured and studied, providing better understanding of the nature of the emitting source.

The second stage of the Cosmic Explorer, CE2, will incorporate new technologies to reduce the thermal noise and quantum noise in the detector. The mirrors in this case will be made of silicon and will again weigh 320 kg. The mirrors will be cryogenic, i.e. they will be cooled to the low temperature of about 123 K ($-150\,°$C). The laser will have a wavelength of 2 μm and the circulating power in the beam will be 2 MW, compared to the 1.4 MW beam power at 1 μm in CE1. CE2 will be able to observe a binary neutron star to a redshift of 26, with signal strength compared to noise being more than double that of CE1. Both CE1 and CE2 will be able to observe black hole binaries, like those detected by aLIGO to large redshifts, extending to the primordial Universe as in the case of the Einstein Telescope.

9.4 Gravitational Wave Detectors in Space

Terrestrial gravitational wave detectors like LIGO have two limitations. First, the seismic, Newtonian and thermal noise increase so much at lower frequencies that it is not possible to detect sources with frequencies below about 1.0 Hz. This is a rich domain to which many types of sources contribute, including supermassive black hole binaries and white dwarf binaries. The other limitation is of arm length: it is not practically possible to increase the arm length beyond a few tens of km, limiting the sensitivity of the detectors, as well as the ability to detect lower frequencies. Both these limitations can be overcome in space. The seismic and Newtonian noise are absent there, and the arm length can be increased using multiple satellites, the only limit being technological. A space-borne gravitational wave detector will of course be a very challenging and expensive task. Nevertheless, given the advantages and the challenges, it will be very exciting and rewarding to build such detectors. We will now describe some such projects expected to be launched in the 2030s, and a pathfinder mission which has already been completed.

9.4.1 LISA

The Laser Interferometric Space Antenna or LISA will be a space-borne gravitational wave observatory, for detecting gravitational waves in the low-frequency range of about 0.1 mHz to about 1.0 Hz. It has been proposed by a consortium of scientists who have been associated with the LISA Pathfinder mission described below, ground-based gravitational wave detectors and other relevant projects. LISA has been accepted by the European Space Agency (ESA). It is expected that LISA will be launched in the mid 2030s and will be operational for at least four years and possibly for some years longer. An artistic impression of a LISA spacecraft is shown in Fig. 9.3.

LISA will consist of three spacecraft orbiting the Sun in a near-equilateral triangle formation. The mean separation between any two spacecraft will be 2.5 million km. The sides of the triangle will act as the arms of an interferometer. The centre of the formation will be on the ecliptic, which is the plane of the Earth's orbit around the Sun. The centre will be at a distance of about 150 million km from the Sun, this being the average Earth-Sun distance. The centre will maintain a distance of between 50–65 million km from the Earth, trailing it by about 20 degrees as it orbits the Sun, as shown in Fig. 9.4. The plane of the triangle makes an angle of about 60 degrees

Fig. 9.3 An artist's impression of one of the LISA spacecraft. Image credit: AEI/MM/exozet

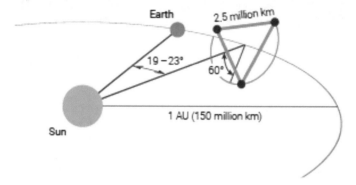

Fig. 9.4 The orbit of LISA around the Sun. The three spacecraft in the mission are configured in a near-equilateral triangle. The centre of the triangle trails the Earth by about 20 degrees. See text for details. Image credit: NASA, An Astro2020 White Paper—The Laser Interferometer Space Antenna

with the ecliptic plane. The formation is maintained throughout the year, but the triangle will appear to rotate about the centre, as shown in Fig. 9.5.

Each spacecraft will have two test masses, each being a cube of size 46 mm made of a gold-platinum alloy, with a mass of 1.96 kg. The test masses will be freely falling in the sense that their motion along the interferometer arms is determined by the gravitational field alone, with any residual disturbance being less than one millionth of a trillionth (10^{-18}) of the acceleration due to gravity on the surface of the Earth. Each test mass is in a vacuum enclosure which reduces electrical disturbances. The enclosure is in communication with the spacecraft, which shields the test masses from outside disturbances, and remains centred on one of the masses using a new technology called the drag-free operation. Each test mass is dedicated to a single interferometry arm, and acts as end point for the arm.

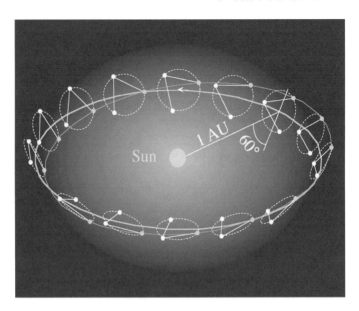

Fig. 9.5 The orbit of LISA around the Sun over a year. The triangle appears to rotate around its centre once per year, so that the plane of the triangle always faces the Sun. Image credit: ESA

In LIGO, the test masses are the suspended mirrors which reflect the powerful laser beam which is incident on the mirror after travelling along the beam tube. Such an arrangement is not possible with the test masses in LISA, because of the great distance the laser beam has to travel from one spacecraft to another. There are no beam tubes connecting the satellites of course, the vacuum of space will allow laser beams to pass without scattering from one satellite to the other. The laser operates at a wavelength of 1064 nm as in LIGO, but it has a power of only 2W. A 30 cm telescope is used in each spacecraft to transmit the laser beam along the interferometer arm, and also to receive the incoming beam. In travelling over the long distance, the beam spreads out due to an effect known as diffraction, as a result of which only a small fraction of the energy, just a few hundred pw reaches the telescope in the receiving spacecraft. This beam is first received by a set of instruments on a platform called an optical bench, and after processing is passed to a test mass, then back to the platform to be returned to the other spacecraft. There is such a platform for each test mass. The complex arrangement is capable of detecting test mass displacements as small as a pm. Given the very long length of the LISA interferometric arms, this provides a strain sensitivity of about 10^{-20}.

Two of the independent interferometers formed from the three arms of the triangular configuration are used to detect gravitational waves. The third interferometer is used to monitor the laser frequency noise in the local laser system. The data processing is done on the ground, using signals transmitted by one of the spacecraft. LISA is an all-sky monitor, so it will be able to detect signals coming from all directions in the sky. It will be able to simultaneously measure the two polarisations that gravitational waves have. Due to the rotation of the triangular configuration and the motion of the configuration around the orbit, it will be possible to observe a gravitational signal which lasts long enough to be observed from different points along the orbit, so that the location in the sky can be determined.

A great advantage of having an interferometer in space is its ability to observe low-frequency gravitational waves. We have seen in Sect. 7.5.5, that at frequencies lower than about 100 Hz, the sensitivity of the aLIGO detector is limited by the rise in thermal, Newtonian and instrumental noise towards lower frequencies. Of these, seismic and Newtonian noise are not relevant in space, as the detector is not placed on the ground. The thermal environment is more stable in space and thermal noise can be controlled. In the LISA frequency band of about 0.1 mHz to about 1 Hz, at lower frequencies the noise is dominated by *acceleration noise*. This disturbs the test masses and is due to a number of reasons like small electrical forces acting on the test masses and the impact of gas molecules on them. At higher frequencies, *position noise* dominates. This includes effects like quantum noise, and the error in the phase of the laser beams while combining them for interference. An additional source of noise is the background due to the combined effect of a large number of weak binary and other sources mentioned in Sect. 5.5. LISA's ability to detect low-frequency gravitational waves is optimised by the great length of the interferometer arms. The larger the length, the more sensitive is the detector to longer wavelengths and therefore lower frequencies. On the Earth, it would be very difficult to have arms significantly longer than the 40 km arms of the Cosmic Explorer.

A rich variety of sources can be found at low frequencies. These include the low frequency gravitational wave emission by the kind of sources already detected by aLIGO and Advanced VIRGO, as well as other sources which are not accessible to those detectors. The first source to be observed by aLIGO, GW150914, was detectable for a very short time, when the two black holes in the binary were moving very rapidly around each other, emitting in the aLIGO band. At earlier stages in its evolution, the black holes would be much farther apart, and would emit low-frequency emission. The source would then have been observable by LISA. Observation of such a source would enable LISA to send an alert to Earth-based detectors that after a certain passage of time, they can expect to see signals emitted from the source in the last stages of in-spiral and during the merger. Better modelling of the source would also be possible because of the ability to detect it over an extended period of time.

White dwarf binaries and ultra-compact binaries of stars would also be emitting at low frequency because of their relatively large orbits and would be observable over a long duration, during which their signal could be added to compensate for its weakness. A very large number of such sources are expected in our galaxy which may not be observable individually but could contribute to a stochastic background which would be observable by LISA. It is expected that there will be binaries of black holes of thousands of Solar masses, known as intermediate mass black holes, and of millions to billions of Solar masses, known as supermassive black holes. Many such binaries will be detectable by LISA over months and even years, providing information about their evolution. Accurate observations of the waveforms of the binaries will allow general relativity to be tested in the regime of very strong gravitational fields. The change in the number and properties of the binaries with their cosmological redshift will provide valuable information about the evolution of the Universe. LISA would also be able to observe gravitational waves from the very early epochs of the Universe and from sources like cosmic strings if they exist.

9.4.2 LISA Pathfinder

The LISA Pathfinder (LPF) was a mission to test some critical technologies needed for the LISA mission, including the accuracy with which the two test masses in each LISA spacecraft could be considered to be in free fall. LPF was launched on 3 December 2015, and placed in orbit around a special point of the Sun-Earth system known as the first Lagrangian point L1, which is at a distance of 1.5 million km from the Earth. The mission was successfully completed and contact with the spacecraft was ended on 17 July 2017, after placing it on a safe orbit around the Sun. LPF was led by ESA, with participation of scientific institutes and the industry, and contribution of a payload from NASA.

The payload with the test masses is the LISA Technology Package (LTP). This has the two test masses freely falling 38 cm apart from each other, each with its vacuum enclosure. There is a high precision interferometer between them for measuring the distance between the test masses and their attitude (orientation). There is also a drag-free control system we mentioned in Sect. 9.4.1 for keeping the spacecraft centred on one of the test masses, and some other instruments. The microthrusters needed for the drag-free control of the spacecraft were provided by NASA as part of a payload package. During the launch and later until the spacecraft reached its orbit around L1, the test masses had to be held in place for protection. A major mission milestone was achieved when the masses were successfully released over a two day period. An artist's impression of the LISA Pathfinder spacecraft is shown in Fig. 9.6.

In the ideal case, the two masses would be in free fall in the gravitational field of the Sun, with a constant distance maintained between the two masses. But the masses are subject to small disturbing forces which produce accelerations and displacements, which act as noise in the process of detecting gravitational waves. The forces include the magnetic and electric forces in the spacecraft, as well as the gravitational force of the spacecraft itself. These have been reduced to the minimum by using non-magnetic materials to the extent possible, by providing shielding against the electrical forces using the enclosure, and by carefully designing the spacecraft to reduce gravitational imbalances. The aim of the LPF was to estimate such noise and to develop the understanding of the forces that cause it. The information will be used to produce accurate free fall by the test masses in LISA.

After a series of experiments, it was established that the test masses were almost motionless relative to each other, and the spurious acceleration between them was estimated to be less than 10^{-16} (one ten millionth of a billionth) of the acceleration due to gravity on the surface of the Earth, over a wide range of frequencies. This performance was significantly better than originally specified.

Fig. 9.6 Artistic impression of the LISA Pathfinder spacecraft. Image credit: ESA

9.4.3 TianQin

TianQin is a space-borne experiment for the detection of gravitational waves in the 0.1–100 mHz band, proposed by the Chinese Academy of Science. The main aim of the mission will be to observe in great detail a single source in the galaxy which is expected to be a strong emitter of gravitational waves. The design of the mission and instruments, the detector and the data analysis will be optimised to the observation of this single *reference source*. The mission is expected to be launched in 2033.

TianQin will have three spacecraft orbiting the Earth at a distance of about 10^5 km, in a formation which will approximately be an equilateral triangle. The three arms of the triangle will act as interferometer arms, each with a length of 10^5 km, similar to the orbital size. This length is considered enough for the planned observation of the reference source. The plane of the triangle will be facing the reference source. The geocentric orbit will enable already tested space technology to be used, reducing the cost of the mission compared to a heliocentric orbit like that of LISA.

As in the case of LISA, freely falling test masses will be present in each spacecraft. The interferometer will measure the change in the separation caused between them by the passage of a gravitational wave. Each test mass will be a cube made of gold-platinum alloy, with each side of 5 cm and mass 2.45 kg. The laser will have a wavelength of 1064 nm with power of 4 W, and the telescopes used to send the laser beam to one of the other spacecraft, and receive a beam from it, will have a diametre of 20 cm. A set of special filters will be used to block sunlight from entering the telescope and a thermal control system will be used to reduce temperature fluctuations. Observations will be limited to two periods of about three months every year, during which the Sun is at large angle to the plane of the triangular configuration. During these periods, the effect of the sunlight entering the telescope is minimised. Each spacecraft will have a disturbance reduction system to reduce the forces that will cause disturbance to the free fall of the test masses. The residual acceleration will have to be less than a millionth of a billionth (10^{-15}) of the acceleration due to gravity at the surface of the Earth. The positional accuracy needed for the mirror is about 1.0 pm.

The reference source which has been tentatively chosen is RX J0806.3+1527, which was first discovered as an X-ray source. X-ray and optical observations have established this source as an ultra-compact binary system with an orbital frequency of `3.11 mHz, which corresponds to an orbital period of 321.5 s and orbital separation of 6.6×10^4 km. The masses of the components are 0.5 and 0.27 Solar masses, respectively. The components should be compact for a binary with such a short period, which corresponds to a short orbital distance. The components are believed to be white dwarfs. The gravitational waves emitted by this white dwarf binary will be detectable by the interferometer with sufficiently strong signal after adding the signals over a three-month observing window.

A pathfinder satellite to test various technologies to be used in TianQin was launched in August 2019. The satellite, called Taiji-1, tested technologies related to the space laser interferometer and drag-free control. Several such pathfinders will be launched for testing before the launch of TianQin around 2033.

9.4.4   DECIGO

The DECi-hertz Interferometer Gravitational wave Observatory (DECIGO) is a planned Japanese space-borne mission to detect gravitational waves in the frequency range of 0.1–10 Hz. The aim is to bridge the gap between the frequency ranges covered by LISA and the present and future ground-based detectors. An advantage of the chosen frequency band is that the stochastic gravitational wave background is low in this region, making it possible to accurately observe individual sources in the band.

DECIGO will be a constellation of four clusters, each cluster consisting of three drag-free spacecraft in an approximately equilateral triangle configuration. The four clusters will be in orbit around the Sun, with two clusters being at nearly the same position. There will be three interferometers per cluster, each using two arms of the triangular configuration. The arm length will be about 1000 km. Each freely falling test mass will be a mirror with diametre 1 m and a mass of 100 kg. The laser source is expected to have a wavelength of 515 nm and power output of 10 W. The sensitivity of the detector is required to be limited only by quantum noise at all frequencies, so the other sources of noise will have to be carefully reduced.

The main mission will be preceded by a pathfinder, DECIGO-B, which will consist of three satellites in a triangular configuration in orbit around the Earth at about 2000 km from the surface. The arm length will be about 100 km, with scaled down specifications for all the components and the sensitivity. The aim of the pathfinder will be to test various technologies and to demonstrate the detection of gravitational waves, establishing the validity of the design.

9.5 Gravitational Wave Detection with Pulsar Timing Arrays

Gravitational waves can be detected through their small effect on the observed periods of pulsars. We have seen in Sect. 5.4.2.1 that many neutron stars have been observed to emit pulses at radio wavelengths. These are fast rotating neutron stars with a high magnetic field. A small part of the radiation they emit, because of the rotation of their magnetic field, is channelled into narrow beams which sweep the sky with the rotation of the neutron star. Each time the beam passes across the field of view of a radio telescope on the Earth, a short burst of radio emission is observed, which over time is seen as a sequence of pulses, with their period being equal to the rotation period of the pulsar.

The rotation of the pulsar remains very constant because it is very massive and its rotation cannot be easily perturbed. So, the period of the pulses too remains constant over long stretches of time, except for occasional glitches. Each observed pulse has a complex structure which can change with time. But when the average of a large number of pulses is taken, the structure is seen to be very smooth and constant. The periods of pulsars generally range over 0.1–10 s, and their magnetic field range over about 10^{11}–10^{13} G, though there are many pulsars outside these ranges too. The energy that a pulsar emits comes from its energy of rotation, so as time passes, the rotation slows down and the rotational period increases. The smaller the magnetic field, the lesser is the rate at which energy is emitted, and the lesser is the slowdown. For an average pulsar, the slowdown is as small as a hundred millionth of a second per year.

There is a special class of pulsars known as *millisecond pulsars*. The first such pulsar to be discovered had a period of 1.56 ms. The fastest among about 200 such pulsars now known has a period of 1.4 ms. A characteristic of millisecond pulsars is their low magnetic field, in the range of about 10^8–10^9 G. Millisecond pulsars are found as members of binary systems as well as single objects.

When a typical pulsar is born, it is spinning rapidly, with a period of about 10 ms, and has a magnetic field of about 10^{12} G or more. This combination of fast rotation and high magnetic field leads to rapid loss of rotational energy by the pulsar, so that the rotational period increases relatively rapidly too, until the pulsar moves into the range for typical pulsars mentioned above. It can remain in the range for 10s to 100s of millions of years. During this long duration, the period of the pulsar increases gradually, and the magnetic field decreases because of various processes, until the pulsar can no longer emit the narrow beams and therefore can no longer be detected.

By the above arguments, the very low magnetic field of millisecond pulsars should make them unobservable, but they can still be detected as pulsars because of their very rapid spin. It is believed that this combination of rapid spin and low magnetic field, which does not occur in typical pulsars, arises because the millisecond pulsars have been *recycled*. They are, or once were, a member of a binary system, in which their companion was a normal star. As the companion evolves, it expands and transfers matter to the neutron star through the process of accretion. The accreted matter carries angular momentum to the neutron star, which spins it up. The neutron stars are old, so they have a low magnetic field, which can be further lowered due to effects produced by the accretion. Once the accretion stops, the neutron star can be observed as a millisecond pulsar with a low magnetic field. During the evolution, the binary system can sometimes be disrupted, in which case a single millisecond pulsar is formed.

The spin of millisecond pulsars decreases very slowly, at a rate which can be as small as 10^{-14} s of slowing down or less per year. Such a millisecond pulsar can be used as a very accurate clock, with the passage of time being indicated by the number of successive pulses counted. But in practice there are errors in the measurement due to various small irregularities in the spin of the pulsar, the orbital motion of the pulsar if it is a binary, errors made during the observational process, effect of the orbital motion of the Earth and the influence of planets and other bodies in the Solar system, errors due to irregularities in the interstellar medium between the pulsar and the Earth, etc. These errors result in differences between the predicted and observed arrival time of pulses, known as *timing residuals*. The errors due to these known effects can be modelled and subtracted from the observed residuals. If all the residuals produced by known effects are properly accounted for in this manner, then the only residuals left are the random fluctuations in the measurements. If systematic residuals beyond expected random fluctuations are observed, those could be because (1) the subtraction of known effects has not been properly done because of errors in the models or data used, or (2) there are other effects which have to be understood and modelled.

Observed timing residuals for two millisecond pulsars are shown in Fig. 9.7. In the figure, the epoch of observation is shown on the horizontal axis, and the residuals in microseconds are shown on the vertical axis. The residuals for the millisecond pulsar B1855+09 in the lower panel show no systematic pattern and only random fluctuations are seen. But in the upper panel for the millisecond pulsar B1937+21, a systematic trend is visible. The deviation is small, being a few microseconds over a decade. This was comparable to the best long term accuracy available from terrestrial clocks in the early 1990s. Such deviations have to be accounted for.

Gravitational waves can contribute to pulsar timing residuals. These waves will affect the structure of space-time in such a way that there is a periodic change in path length between the pulsar and the Earth. This is similar to the change in arm lengths of the aLIGO detectors when a gravitational wave passes through, which is detected using a Michelson interferometer. In the case of a pulsar, it is the pulse

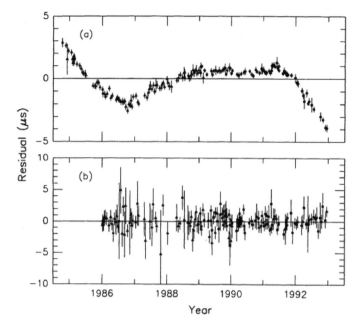

Fig. 9.7 Timing residuals for two millisecond pulsars, B1937+21 and B 1855+09. Image credit: V. M. Kaspi, J. H. Taylor and M. F. Ryba, Astrophysical Journal, **428**, 713, 1994

arrival time which changes, leaving a mark on the residuals. The gravitational waves produce a change in the gravitational field at the pulsar, leading to a *pulsar term*, and at the Earth leading to an *Earth term*. Both these terms contribute to the delay in time of arrival of a pulse, leading to residuals. The timing of millisecond pulsars is needed to detect these small residuals, since millisecond pulsar periods are extremely steady as described above. The residuals produced by various kinds of gravitational wave sources can be modelled, and the observed residuals can be examined for these signatures.

The frequency of the gravitational waves which can be detected in this manner is limited to the range of 10^{-9}–10^{-7} Hz due to the present observational constraints. Such very low frequencies are produced by supermassive black hole binaries with component mass greater than about 10^8 Solar mass, and orbital periods ranging from a few years to tens of years. The wavelength of such radiation is in light years and the timing residuals produced are in the range of about 10 ns. Supermassive black holes which are too faint to be detected individually will contribute to a stochastic background, which can again be studied at the low frequencies. Other sources which could possibly be detected are cosmic strings and gravitational waves produced in the inflationary stage of the early Universe.

The accuracy with which pulsar timing residual can be measured can be improved significantly using pulsar timing arrays (PTA). These are a collection of a number of millisecond pulsars located in different directions in the sky. Each pulsar is individually observed to obtain its timing residuals. The pulsar term is different for each pulsar and the corresponding residuals are uncorrelated. But the Earth term produces correlated patterns of residuals among the pulsars. These correlations depend on the angle between the lines of sight from the Earth to the source of gravitational waves and to the pulsar. Careful analysis of the residuals from the millisecond pulsars in the array can lead to detection of gravitational waves.

The pulsar timing work is carried out through a collaboration known as the International Pulsar Timing Array, which is made up of smaller PTA collaborations. A number of radio telescopes located in different countries are used for timing observations of a large number of pulsars. So far, no definitive gravitational wave contribution to the residuals has been observed, but useful limits have been obtained. It is expected that over the present decade, as techniques become more sensitive, new pulsars are added to the array and observations are conducted over a longer period of time, the detection of gravitational waves using PTAs will be made. A highly sensitive new radio telescope, the Square Kilometre Array (SKA), will be ready for observations over the present decade. It will be able to carry out timing observations of hundreds of pulsars with accuracy better than a microsecond, making it easier to detect gravitational waves.

Further Reading

We provide below a short list of books on various topics covered by us. The books listed are mostly widely used text books at the undergraduate level and beyond, with which we are familiar. The books require some knowledge of physics and mathematics, and bridge the gap between our qualitative treatment and more advanced, specialised books. We have listed just a few examples from the many excellent books which are available on each topic.

Books on Electromagnetism and Special relativity

Feynman, Richard P.—The Feynman Lectures on Physics 2, 2nd Edition, 2005. Addison-Wesley.
Griffith, David—Introduction to Electrodynamics, 4th Edition, 2012. Addison-Wesley.
Krane, Kenneth, Halliday, David & Resnick, Robert—Physics 2, 2nd Edition, Wiley.
Morin, David J.—Special relativity: For the Enthusiastic Beginner, 2019. Create Space Independent Publishing Platform.
Taylor, Edwin, F. & Wheeler, John Archibald—Space-Time Physics: Introduction to Special Relativity.

Books on Astrophysics

Longair, Malcolm S.—High Energy Astrophysics, 3rd Edition, 2011. Cambridge University Press.
Prialnik, Dina—An Introduction to the Theory of Stellar Structure and Evolution, 2nd Edition, 2011. Cambridge University Press.
Raine, Derek & Thomas, Edwin—Black Holes: A Student Text, 3rd Edition, 2014. ICP.
Shapiro, Stuart L. & Teukolsky, Paul A.—Black Holes, White Dwarfs and Neutron Stars: The Physics of Compact Objects, 1983. Wiley-VCH.
Srinivasan, Ganesan—What are the Stars?, 2011. Springer.
Srinivasan, Ganesan—Life and Death of Stars, 2014. Springer.

© The Editor(s) (if applicable) and The Author(s), under exclusive license
to Springer Nature Singapore Pte Ltd. 2020
A. Kembhavi and P. Khare, *Gravitational Waves*,
https://doi.org/10.1007/978-981-15-5709-5

Books on General relativity and Gravitational Theory and Gravitational Waves

Anderson, Niels—Gravitational-Wave Astronomy: Exploring the Drak Side of the Universe, 2019. Oxford University Press.
Carroll , Sen, M.—Spacetime and Geometry, 2019. Cambridge University Press.
Hartle, James, B.—Gravity: An Introduction to Einstein's general Relativity, 2014. Pearson.
Misner, Charles, W., Thorne, Kip, S. & Wheeler, John Archibald—Gravitation, 2017. Princeton University Press.
Narlikar, Jayant, V.—The Lighter Side of Gravity, 1996. Cambridge University Press.
Narlikar, Jayant, V.—Introduction to Relativity, 2006. Cambridge University Press
Schutz, Bernard—A First Course in General Relativity, 2nd Edition, 2012. Cambridge University Press.
Schutz, Bernard—Gravity from the ground up, 2003. Cambridge University Press.

Useful Websites for Gravitational Wave Detections

The website ligo.org maintained by the LIGO Scientific Collaboration and the website ligo.caltech.edu provide a wealth of information about all aspects of gravitational wave physics, gravitational wave observatories and gravitational wave detections. A list and description of all gravitational wave candidate events is available on the GraceDB—Gravitational-Wave Candidate Event Database at gracedb.ligo.org.

Glossary

Absolute temperature—Temperature measured on the Kelvin scale.

Absorption lines—Dark lines in the spectrum of a star produced by absorption of radiation, travelling from the interior of the star to the outside by matter present along its path.

Advanced LIGO—An upgraded version of the initial LIGO detector. Its sensitivity is about 10 times that of initial LIGO.

aLIGO—Advanced LIGO.

Angular diameter—Angle subtended by two diametrically opposite points of an astronomical object at the observer's eye.

Apparent brightness—Energy received from the star by the observer per unit area per second.

Binary system (Binary stars)—A pair of stars orbiting around each other.

Black body—An object that absorbs all electromagnetic radiation falling on it. The nature of the radiation emitted by such a body depends only on its temperature.

Black hole—An object whose gravity is so strong that not even electromagnetic radiation can escape from it.

Brownian noise—Disturbances in the mirrors of gravitational wave detectors caused by the heating of mirrors and their suspensions, and the mechanical loss of the coatings of mirrors caused by the laser beams of the interferometers.

Cepheid variable—A type of star whose luminosity varies periodically with a characteristic pattern due to the periodic oscillations of the surface of the star.

Chandrasekhar limit—The maximum mass that a white dwarf can have.

Chirp mass—A mass associated with a binary system which is spiralling-in due to the emission of gravitational waves. The chirp mass depends on the mass of the two components of the binary, and can be determined from observable quantities.

Chromosphere—The second outermost layer of the Sun, located between the photosphere and the corona. It is several thousand kilometres thick.

Compact sources—Cosmic objects which are remnants produced at the end stages of a star's life. They have a radius much smaller than the radius of a normal star and have very high density. Compact objects include white dwarfs, neutron stars and black holes.

Corona—The outermost layer of the Sun reaching temperatures of about a million Kelvin and extending to a few Solar radii.

Cosmic Explorer—A laser interferometric gravitational wave detector which is being planned. It will be developed in two stages, will have 40 km long arms and will be located in the USA.

Cosmological redshift—Redshift caused by the expansion of the universe.

DECIGO—The DECi-hertz Interferometer Gravitational-wave Observatory is a planned Japanese space-borne mission to detect gravitational waves in the frequency range of 0.1–10 Hz.

Degeneracy pressure—Pressure due to high density electron or neutron gas caused by quantum mechanical effects.

Degenerate gas—High density electron or neutron gas where quantum mechanical effects are important.

Einstein's equations—Equations derived by Albert Einstein which determine the gravitational field for given matter and energy distribution.

Einstein's ring—A ring shaped image of an astronomical source caused due to gravitational lensing by an intervening galaxy or galactic cluster along the line of sight.

Einstein telescope—A laser interferometric gravitational wave detector being planned, which will have three 10 kilometre long arms in a triangular configuration and will be located in Europe.

Electromagnetic waves—Waves having periodically varying electric and magnetic fields.

Equivalence principle—The statement that at any point in a gravitational field, all bodies, irrespective of their mass, would have the same acceleration.

Escape velocity—The minimum velocity with which a mass has to be thrown outward from the surface of an object for it to escape the gravitational field of the object.

Ether—Hypothetical medium which was believed to be necessary for propagation of electromagnetic waves. This belief was abandoned after Einstein developed his special theory of relativity.

Event horizon—An imaginary surface surrounding a black hole from which no signal can emerge to the outside world.

Fraunhofer's lines—Dark absorption lines seen in the Solar spectrum.

Free particle—A particle not influenced by any force.

Galaxy—A gravitationally bound system of stars, stellar remnants, interstellar gas, dust and dark matter.

Gamma rays—Most energetic part of electromagnetic spectrum having wavelength smaller than 0.01 nanometre.

Gamma-ray burst—Cosmic bursts of highly energetic gamma rays lasting from less than a second to several minutes.

GCN—Gamma-ray Coordinates Network, which is a portal for discoveries and observations of astronomical transients, including those at electromagnetic wavelengths and gravitational-wave, cosmic-ray and neutrino events.

General theory of relativity—Theory of gravity given by Einstein in 1915 in which gravity is manifested as curvature of space-time.

Gravitational lensing—Distortion and magnification of light caused by the bending of rays of light by a gravitational field. This was predicted by Einstein's general theory of relativity.

Gravitational potential energy—The stored energy due to the gravitational interaction between two masses. Conventionally this is taken to be negative. The greater the gravitational force between two masses, the more negative is the energy between them.

Gravitational redshift—Increase in the wavelength of electromagnetic radiation while moving from a region of strong gravitational field to a region of weak field.

Gravitational waves—Periodic disturbances in the curvature of spacetime, i.e. in the gravitational field, generated by accelerated masses, that propagate as waves outward from their source at the speed of light.

IndIGO—Indian Initiatives in Gravitational Wave Observations, which was established in 2009 as a consortium of interested researchers to work together to participate in developing experimental facilities for gravitational wave detection, and in their observations and data analysis.

iLIGO—Initial LIGO.

Infrared radiation—Electromagnetic radiation with wavelengths in the range of about 700 nanometre to about 1 millimetre.

Initial LIGO—The LIGO detector which was functioning till 2015. This has been superseded by Advanced LIGO.

Interference—Phenomenon in which two waves superpose to form a resultant wave of greater, lower, or the same amplitude. The pattern in the intensity produced due to interference is known as an interference pattern.

KAGRA—The Kamioka Gravitational Wave Detector, a laser interferometer located underground, in the Kamioka mine in Japan.

Kepler's laws—Three laws given by Johannes Kepler in the 17th Century describing the motion of planets around the Sun.

Kerr black hole—A rotating black hole. The solution of Einstein's equations which describes such a blackhole was discovered by Roy Kerr.

Kerr-Newman black hole—A rotating black hole with electric charge. The solution of Einstein's equations which describes such a blackhole was discovered Ezra T. Newman.

Laser—A laser is a device which emits light through a process known as the stimulated emission of radiation. The emitted radiation is coherent, has a specific wavelength and spreads very little as it travels forward. A laser beam is emitted by a laser device.

LIGO—Laser Interferometric Gravitational Wave Observatory. The Observatory consists of a LIGO gravitational wave detector located in Hanford, Washington State and another at Livingston, Louisiana.

LIGO A+ —An upgraded version of advanced LIGO to be ready in 2023.

LIGO-India—An Advanced LIGO detector to be installed in India.

LIGO Scientific Collaboration (LSC)—A collaboration of scientists who carry out scientific and technical work related to LIGO.

LIGO Voyager—The LIGO detector with the ultimate sensitivity which can be reached with the present LIGO infrastructure.

LISA—The Laser Interferometric Space Antenna, a future space-borne gravitational wave observatory.

LISA Pathfinder—A space mission launched to test technologies needed for LISA.

Luminosity—The total energy emitted by a star per second.

Mass gap—A mass range from about 2 to about 3 times the Solar mass. At the present time there are no known black holes with mass in this range.

Maxwell's equations—A set of equations to determine the electric and magnetic fields for given charge and current distributions named after the physicist James Clerk Maxwell.

Michelson-Morley experiment—An experiment based on the Michelson interferometer which was performed in 1887 by A. A. Michelson and E. W. Morley to detect the presence of the ether.

Millisecond pulsar—A pulsar with rotation period less than about 40 milliseconds.

Multi-messenger astronomy—Astronomy performed with different messengers like electromagnetic radiation, gravitational waves, cosmic rays and neutrinos.

Neutron star—A compact object made mostly of neutrons in which neutron degeneracy pressure balances gravity. It is the end stage of intermediate mass stars.

Newtonian noise—Disturbances in the mirrors of gravitational wave detectors caused by small changes in the density of the ground due to seismic disturbances. This is also known as gravity gradient noise.

Newton's law of gravitation—The law that gives the gravitational force between two objects in terms of their masses and the distance between them.

Orbital motion—The motion of a particle or a body moving under the influence of a force due to another body, like the motion of a planet around the Sun which exerts an attractive gravitational force on it.

Parallax—The shift in the apparent position of an object, relative to the background, due to a change in the position of the observer.

Perihelion—The point along the orbit of a planet, asteroid or comet that is nearest to the Sun.

Photon—Smallest discreet amount of energy of electromagnetic radiation. It is the particle form of electromagnetic waves. The energy of a photon is proportional to the frequency of the radiation.

Photon shot noise—The random fluctuation with time in the number of photons in a beam of radiation of assumed constant brightness.

Piezoelectric crystal—A material which generates electric charge when pressure or strain is applied to it. A piezoelectric transducer is a device which uses the piezoelectric effect to measure changes in pressure or strain.

Planck's constant—A fundamental constant in quantum physics first introduced by Max Planck. The energy of a photon is equal to the product of the Planck's constant and the frequency of the radiation.

Polarisation—The process of converting unpolarised waves to polarised waves.

Polarised Waves—Transverse waves in which the oscillations are confined to a fixed direction perpendicular to the direction of propagation are said to be linearly polarised waves. When the direction of oscillation rotates in the plane transverse to the direction of motion, the wave is said to be circularly or elliptically polarised.

Precession—A slow change in the orientation of the orbit of one body around another, like the orbit of a planet around the Sun.

Pulsar—Highly magnetized rotating neutron star which emits electromagnetic radiation and particles in two narrow oppositely directed beams along its magnetic axis.

Pulsar Timing Array—A collection of millisecond pulsars whose rotation period and the rate at which it changes are very accurately known. Such an array can be used to detect very low frequency gravitational waves.

Quantum mechanics—The branch of physics applicable to the motion and interactions of atomic and subatomic particles.

Quantum noise—Photon shot noise.

Quasar—A very compact cosmic object which emits an exceptionally large amount of energy over the entire range of the electromagnetic spectrum. The emission is believed to be due to matter falling on to a supermassive black hole at the centre of large galaxies. At optical wavelengths a quasar image can have the appearance of a star, so the first quasars to be observed were termed as quasi-stellar objects and hence their present name. Quasars are amongst the most distant objects known in the Universe.

Radial velocity—The velocity of an object along the line of sight of an observer.

Random noise—The small changes with time in a measured signal which are produced due to random effects.

Redshift—Increase in the wavelength of electromagnetic radiation during its passage from a cosmic object to an observer. The increase can occur due to the motion of the emitter away from the observer, which is known as the Doppler effect, or can be due to gravitational effects or cosmic expansion.

Ringdown—The phase through which two merged compact objects settle down to a single black hole with mass and spin as its only two properties.

Schwarzschild solution—Solution of Einstein's equations for a massive, non-rotating point source. It was discovered by Karl Schwarzschild.

Schwarzschild radius—Distance of the event horizon from the centre of a non-rotating black hole which is associated with the Schwarzschild solution.

Seismic noise—Disturbances in the positions of mirrors of gravitational wave detectors caused by seismic disturbances including those caused by human beings, winds and tidal motions in the ground caused by the Sun and the Moon.

Special theory of relativity—A theory of the relationship between space and time developed by Albert Einstein.

Spectrum—The spread of radiation across a range of wavelengths. For electromagnetic waves this is known as the electromagnetic spectrum. A spectrum can have a continuous spread of energy over a range of wavelengths as well as discrete features at specific wavelengths.

Spiral-in—The motion of the two components of a binary towards each other along a spiral path due to the loss of potential energy before their merging. During the spiral-in the members of the binary move closer to each other with ever increasing speeds.

Strain—The ratio of the change in the arm length of the LIGO detector, caused by the passage of a gravitational wave, to the length of the arm.

Strain noise—The strain produced due to various noise sources.

Supermassive black holes—Black holes with masses from hundreds of thousands to billions of Solar masses. These are believed to be present at the centres of most large galaxies.

Supernova explosion—Powerful and luminous explosion of a star towards the end of its lifecycle resulting in a neutron star or a black hole.

Thermal energy—Internal energy possessed by an object in the form of heat.

Thermal noise—Brownian noise.

TianQin—A space-borne experiment for the detection of gravitational waves in the 0.1–100 mHz band being built by the Chinese Academy of Science.

Ultraviolet radiation—Electromagnetic radiation with wavelengths between in the range of about 100 to about 400 nanometres.

VIRGO—A laser interferometer for gravitational wave detection located near Pisa in Italy.

Visible radiation—Electromagnetic waves to which our eyes are sensitive. This has wavelength of about 380 to about 740 nanometres.

Wave equation—An equation describing the propagation of a wave in space and time.

White dwarf—The end stage of low mass stars. The gravity of the white dwarf is balanced by electron degeneracy pressure to provide it a stable structure.

X-rays—High energy electromagnetic radiation with wavelength in the range of about 0.001 to about 10 nanometres.

X-ray binary—An X-ray emitting binary system consisting of a compact object and a normal star. The X-rays are emitted when some matter from the normal star is transferred to the compact object, with the matter becoming hot enough to emit X-rays.

Index

Printed in the United States
By Bookmasters